22nd Topical Conference on Refinery Processing 2019

Topical Conference at the 2019 AIChE Spring Meeting
and 15th Global Congress on Process Safety

New Orleans, Louisiana, USA
31 March - 4 April 2019

ISBN: 978-1-5108-8677-3

Printed from e-media with permission by:

Curran Associates, Inc.
57 Morehouse Lane
Red Hook, NY 12571

Some format issues inherent in the e-media version may also appear in this print version.

Printed by Curran Associates, Inc. (2019)

For permission requests, please contact AIChE
at the address below.

AIChE
120 Wall Street, FL 23
New York, NY 10005-4020

Phone: (800) 242-4363
Fax: (203) 775-5177

www.aiche.org

Additional copies of this publication are available from:

Curran Associates, Inc.
57 Morehouse Lane
Red Hook, NY 12571 USA
Phone: 845-758-0400
Fax: 845-758-2633
Email: curran@proceedings.com
Web: www.proceedings.com

22nd Topical Conference on Refinery Processing 2019

Topical Conference at the 2019 AIChE Spring Meeting
and 15th Global Congress on Process Safety

New Orleans, Louisiana, USA
31 March - 4 April 2019

AIChE
The Global Home of Chemical Engineers

TABLE OF CONTENTS

549955 Heat Exchanger Fouling and Cleaning Analysis for Complex Network

Heat exchanger fouling has a direct impact on plant profitability. To combat fouling, heat exchangers must be periodically removed from service and cleaned. Frequent cleaning substantially increases cleaning expenditures. However, cleaning too infrequently has the potential to increase energy consumption due to limited heat transfer capabilities of the fouled heat exchangers. In both cases, operating cost goes up.

Many refineries conduct cleaning on a pre-determined schedule. Some only perform cleaning when operating problems, such as hydraulic restrictions or unacceptably high rundown temperatures, become apparent. Others do periodic monitoring of individual heat exchanger performance to identify the most badly fouled heat exchangers. However, a heat exchanger is not necessarily worth cleaning simply because it is heavily fouled. In case of a crude unit, heat exchangers at the front-end of the preheat train have very limited impact on the temperature of the crude oil by the time it reaches the atmospheric heater, as a loss of heat recovery in an upstream heat exchanger is usually offset by increases in heat recovery in downstream heat exchangers. Hence, in addition to individual heat exchanger monitoring, it is also necessary to analyze the knock-on effects throughout the heat exchanger network.

This paper presents heat exchanger fouling and cleaning analysis of a crude preheat train with the objective of mitigating fouling effects through improved cleaning schedules. In this analysis, rigorous heat exchanger network simulation of a crude preheat train at various fouling conditions is performed in UniSim to account for heat exchanger interactions and identify critical heat exchangers in terms of fouling effects that need to be monitored closely. In addition, UniSim simulation results are coupled with economic factors, other asset conditions and operational scenarios to optimize cleaning schedules resulting in overall economic savings (OPEX reduction). The findings from this analysis also provide valuable insights for maintenance scheduling, turnaround planning, and process improvement.

550592 Performance Evaluation of Alternative Technologies for Fouling Mitigation in Refinery Pre-Heat Trains

While heat exchanger fouling management tools can be effectively used in refineries to minimise the impact of deposition from crude oils, it is desirable to mitigate fouling by design. This has been a challenge for many years in part because of the lack of understanding of the fundamental mechanisms leading to deposition and in part for the intrinsic limitations of the methodologies traditionally used for heat exchanger design. Moreover, opportunities to use non-conventional mitigation technologies (e.g. helical baffles, welded plates, tube inserts) are often missed out, given the challenges in evaluating their performance with respect to traditional shell-and-tube heat exchangers.

In this paper, a methodology to evaluate the performance of different technologies for fouling mitigation with plant data is demonstrated. The hot end of a refinery preheat train is analysed to identify for retrofit opportunities to mitigate fouling. Several retrofit options are proposed for the heat exchanger with greater potential for improvement. By using fouling predictive models, previously fitted and validated using plant measurements, it is possible to re-run history and assess how alternative designs would have performed. The evaluation includes the evolution of fouling, the time-variation of the thermal and hydraulic performance of the heat exchanger retrofitted, the network interactions, and, ultimately, the economic savings that would have been achieved if the alternative technologies had been placed instead of the original one. The methodology is shown to provide a quantitative assessment of the benefits and trade-offs involved in the retrofit, de-risking the selection of non-conventional technology.

Proceedings 544039

Crude Unit Fouling: Consequences and Costs

Andrew W. Sloley

Advisian (WorleyParsons Group)
Energy Center II
575 N Dairy Ashford St
Houston, TX 77079

Prepared for Presentation at the
American Institute of Chemical Engineers
Spring Meeting
New Orleans, LA, March 31 – April 4, 2019

UNPUBLISHED

1. Introduction

Most crude unit preheat trains use shell-and-tube exchangers for their heat recovery. Heat exchanger fouling adds considerable cost to both construction and operation of crude units. Higher fouling shifts optimum heat recovery investment to higher capital and lower energy recovery. The industry has taken three approaches to account for fouling in heat exchanger design:

- Fouling factors
- Over-surface
- Cleanliness factors

Fouling factors add a resistance term to the heat-transfer calculations. This factor can be applied to either or both sides of the service. Using a fouling factor is the most common industrial practice. It is used extensively to design exchangers, evaluate performance, and determine exchanger cleaning schedules.

The second approach uses an over-surface specification. The exchanger is designed as if it were clean, then extra area is added. This approach has the benefit of being explicit in the cost of the overdesign but is much more difficult to use in optimizing cleaning schedules of existing equipment.

The cleanliness factor is a de-rating factor applied to the overall heat-transfer coefficient. It functions similarly to an over-surface margin specified when designing new equipment. The cleanliness factor approach is most often used in the design of surface condensers in the power industry. In that service, overall heat-transfer coefficients are high and nearly all the fouling is on the cooling water (tube) side of the exchangers.

Spirited discussion continues on which method is best for the design and evaluation of crude unit exchangers. The exchangers are well known for severe and persistent fouling, a problem that has defeated attempts at solution for decades. Using too low a fouling level leads to units that cannot meet energy recovery, capacity, or quality targets. Using too high a fouling level leads to excessive investment in heat exchangers, and may have little-to-no performance improvement if done incorrectly.

This paper examines both the fouling-factor approach and the over-surface approach and compares them. Data are presented from four operating crude units. All four of the units exclusively use shell-and-tube exchangers to recover energy for heating the crude oil feed. Previous work [1] based on a literature review identified reasonable fouling factors for commercial units, even though fouling may vary dramatically from unit to unit and from exchanger to exchanger within the same unit.

This study answers the following questions:

A. What is the relationship (if any) between fouling factors and an equivalent over-surface percent?
B. What is the heat recovery cost of actual versus TEMA[a] fouling factors?
C. What is a realistic over-surface percent or fouling factor for a crude unit?

2. Factors Affecting Fouling

Regardless of the method used to account for, or alleviate, crude unit fouling, fouling in crude units has been a persistent problem for decades. One of the author's first experiences in his first year of working was to review a colleague's progress in predicting/preventing fouling of crude unit exchangers. There hadn't been any progress for years. Every idea and theory had failed.

The fundamental issue is that crude fouling is complex. It arises from a combination of the following factors:

- Crudes and how their blending properties interact, leading to colloidal and composition interactions
- Crude history including blending order and exposure to oxygen and other chemicals
- Desalting and dehydration performance in the tank farm and in the crude unit
- Composition and structure of the residual mineral content
- Fouling deposit aging, crude deposit reactions, and other factors
- Shear stress (velocity) in the exchanger
- Exchanger configuration including fluid allocation, type, and dead spots in the flow patterns
- Differences between surface, film, and bulk temperatures, and differences in properties arising from them
- Differences in the multiple fouling mechanisms present including asphaltene precipitation, carbonate formation, wax deposition, chlorides and other salt deposition, contained solids deposition, and fouling-corrosion interaction

Along with this comes the difficulty in gathering high-quality data, and uncertainty in correlation accuracy. In operating plants it is impossible to separate the effects of fouling on one side of the exchanger from the effects of fouling on the other side. Also, operating plants may suffer from incomplete exchanger cleaning and exchanger damage. Both of these confuse data analysis.

In the industry as a whole, progress has been made. The progress has been by small, cumulative gains in exchanger design, chemical additives, desalting improvement, and crude blending knowledge. However, large step changes that would prevent energy and capacity losses due to fouling have eluded the industry.

[a] TEMA, Tubular Exchanger Manufacturers Association

3. Fouling Factors

3.1. What are Fouling Factors?

Fouling factors add a thermal resistance to the overall heat-transfer calculation. Equation 1 shows the overall heat-transfer coefficient (U) as a function of the tubes' outside heat-transfer coefficient (h_o), resistance to heat transfer of the outside tube surface (r_o), resistance of the wall (r_w), resistance of the inside tube surface (r_i), inside heat-transfer coefficient (h_i), and surface areas of the outside of the tube (A_o) and inside of the tube (A_i). A fouling factor is used to show the thermal resistance to fouling, giving a value to either r_o or r_i.

Equation 1

$$U = \frac{1}{\left[\frac{1}{h_o} + r_o + r_w + r_i \frac{A_o}{A_i} + \frac{1}{h_i}\frac{A_0}{A_i}\right]}$$

In design, fouling should represent fouling alone. The fouling factor should not be used to substitute for uncertainties in correlations and physical property predictions.

The best source of information on fouling factors should be experience with similar exchangers on similar feeds. However, at the design level this information is rarely available.

TEMA publishes 'Recommended Good Practice' values in their standards [2]. With few updates, these values are mostly the same as they were when originally published in 1941. No documentation exists supporting the original definition of the fouling factors. Based on a summary of discussion with early members of TEMA, the conclusion is that the original values were a *"consensus drawn from many years of practical experience"* [3].

TEMA factors include values for both salted crude (containing produced water with soluble salts), and dry crude. Fouling occurs on both sides of exchangers. TEMA includes factors for both the crude and the heat recovery streams (crude unit products and heat removal streams). The analysis here looks at a total fouling factor of tubeside fouling plus shellside fouling. Based on TEMA, most crude exchangers should be designed with a total fouling factor between 0.004 and 0.007 ft^2-°F-hr/BTU (0.0007-0.0012 m^2-°C/W). For heat integration with residue streams, TEMA recommends total fouling values of up to 0.014 ft^2-°F-hr (0.0025 m^2-°C/W).

3.2. Advantages of Using Fouling Factors

Fouling factors have long been used in the industry. Whether they are accurate or not, the values published by TEMA find common acceptance for design purposes. They also have a fundamental place in the overall heat transfer in Equation 1. How the fouling resistance shows up is clear and transparent. The use of fouling factors allows heat-transfer engineers to optimize exchanger design for specific fouling services. Table 1 shows common parameters that may be used to optimize the exchanger.

Table 1 Exchanger Optimization Parameters

Shell-Side Optimization Features	Tube-Side Optimization Features
Which fluid in shell (fluid allocation)	Which fluid in tube (fluid allocation)
Configuration	Tube passes
Diameter, length	Tube diameter
Baffle type, cut, spacing	Tube length
Tube spacing	Tube count
Tube pattern	Pressure drop
Pressure drop	

TEMA fouling factors are accepted by many as de-facto standards. In recent years, other sources [4] have given examples of fouling factors in specific applications, but little published work gives overall recommendations.

For monitoring, most approaches (and commercial software) calculate a fouling factor based on measured flow rates and temperatures, calculated physical properties, exchanger configuration information, and a selected method for rigorous evaluation of the heat exchanger. The difference between expected clean performance and actual measured performance is balanced by finding the fouling factor that accounts for the difference.

In monitoring, the fouling factor combines fouling, data errors, calculation inaccuracy, problems due to exchanger damage, and any other contributing factors. Because of this, nominally clean exchangers may give results that imply extensive fouling. In other cases, exchangers may even have negative fouling factors.

3.3. Disadvantages of Using Fouling Factors

Designing with fouling factors has three major disadvantages.

First, the fouling factor can obscure how much extra surface area is added to the exchanger to maintain performance when fouled. Often, the cost implications of fouling factors are not well understood. This affects both the design of specific exchangers as well as the overall design and optimization of entire preheat trains.

Second, when fouling factors are high, exchanger areas become large. Large exchangers tend to have relatively low fluid velocities at the optimum design point. Extensive industry experience shows that low fluid velocity (and thus low shear stress) increases fouling rates [5]. The exact relationship can be disputed, but the trends are very clear.

Third, fouling factors are fixed. Fouling is dynamic. In some cases, reliable data have shown that under some conditions for some exchangers, there are periods when fouling decreases. Also, specific fouling values are based on assumptions about cleaning schedules and practices—which are unknown when designing the unit.

3.4. Unimportant Considerations

A fourth criticism is sometimes leveled at fouling factors, namely that their values are too low. That may be true, but this criticism targets data adequacy, not a fundamental fault with the method or its application.

On the other hand, an advantage often quoted for fouling factors is that they explicitly allow designs to account for differences in fouling tendencies of different streams. This has little real value for design, for two reasons. In most crude exchangers, the dominant fouling is on the crude side. Also, when fouling factors are derived from operating plant data, the overall value is split arbitrarily and allocated to one side of the exchanger or the other. Stream fouling factors are thus little more than educated guesses.

3.5. Reasonable Fouling Factors

Defining a reasonable fouling factor requires looking at two things: first, how much fouling happens, and second, is the analysis looking at a single exchanger or an entire preheat train?

As previously noted, using TEMA values will generally result in crude exchangers with total fouling factors from 0.004 to 0.007 ft^2-°F-hr/BTU (0.0007-0.0012 m^2-°C/W).

Based on a late 1980's survey [3], about 70% of new crude preheat exchangers had a specified total fouling factor of 0.004 to 0.006 ft^2-°F-hr/BTU (0.0007-0.0011 m^2-°C/W). However, some exchangers were designed with fouling factors specified as high as 0.030 ft^2-°F-hr/BTU (0.0053 m^2-°C/W), five to eight times the normally accepted value.

Recent review of published data by the author examined 139 exchangers from multiple units [1]. These data are mainly from 1990-2015 and are biased toward problem cases. Fouling followed a bimodal distribution, with one set of low-fouling (normal) exchangers and another set of high-fouling (problem) exchangers. The important parts of this analysis show:

- For the problem exchangers, the mean fouling level when nominally clean at startup exceeded the TEMA values.
- The dominant pattern of fouling in crude exchangers over time is a linear increase, though the fouling can be subject to changes in slope, or display sudden steps.
- For the low-fouling exchangers, the fouling factor seen at end-of-run (or just before cleaning) was 2.4-2.8 times the TEMA values. Forty-nine exchangers were in this group.
- For the high-fouling exchangers, the fouling factor seen at end-of-run (or just before cleaning) was 6.5-8.1 times the TEMA values.
- No correlation was seen for overall fouling factor versus the nature of the stream on the non-crude side of the exchanger.

What constitutes a reasonable fouling factor depends upon the consequences of using the wrong factor. When looking at an entire preheat train, not every exchanger will be at maximum fouling at the same time. Some exchangers will be fouled, some will have just been cleaned, some will be fouling faster than average, some will be fouling slower than average. Additionally, heat not recovered in one exchanger may be recovered by a downstream exchanger.

When looking at a critical service with a single exchanger, the fouling factor should meet the expected end-of-run conditions for that exchanger. For example, in Figure 1, exchanger service EX15A-C is the only exchanger in the kerosene pumparound circuit. If this exchanger transfers less heat than expected, all the lost heat goes to the tower condensers. The increased overhead duty could cause the tower to pressure up (condensers are now too small) or the tower to flood (too high a vapor load in the tower).

The final exchangers in the preheat train are most critical to the overall train performance. There are fewer, or no, exchangers downstream to compensate if they have poor heat recovery. In cases where the downstream heater may limit operation, these should be considered critical service exchangers.

Based on the previous work the author recommends the following fouling factors for design:

- Normal services – 0.0075 ft^2-°F-hr/BTU (0.0013 m^2-°C/W). This is 50% of the end-of-run or just-before-cleaning value for the heat exchanger.
- Critical services – 0.015-0.020 ft^2-°F-hr/BTU (0.0026-0.0035 m^2-°C/W)
- Final services – 0.015--0.020 ft^2-°F-hr/BTU (0.0026-0.0035 m^2-°C/W)

These factors are based on observed plant performance at typical 180-day to 250-day cleaning intervals.

4. Over-surface Factors

Over-surface specifies that a set fraction of extra area should be added to a heat exchanger. For design, the advantage of this approach is that it makes explicit the area (and cost) added to compensate for fouling. When using an over-surface approach, the design engineer should specify whether to add area by increasing diameter or by increasing length.

The preferred approach would be to add exchanger length. Pressure drop limits should be relaxed when factoring in extra surface. The hydraulics in the overall heat train should have enough allowance to account for the extra pressure drop.

In real units (rather than paper studies) most plants only allow specific standardized exchanger lengths for heat exchangers. This makes unit layout simpler and reduces costs by standardizing replacement tubes. If the specifications are set so that one standard length replaces another standard length (i.e. 16-ft long exchanger plus 25% over-surface gives a 20-ft long exchanger) this approach works well.

The alternative is to design exchangers to maintain a specified minimum fluid velocity when the over-surface is added. This allows the designer some flexibility in fitting an exchanger within standard dimensions. Again, pressure drop specifications should be flexible enough to account for the extra pressure drop required to meet the velocity specification.

The greatest problem with specifying an overdesign is that few engineers have any feel for how much is needed. When discussed, over-design values of 25-30% are often mentioned. These values are usually arbitrary and not based on sound economic analysis or verified data.

5. Operating Unit Comparisons

Data from test runs of four operating units have been analyzed to examine the consequences of fouling. None of these data were included in the individual exchangers reviewed in the previous work [1]. The units were chosen for the following major reasons:

Unit 1: A unit configured with heat integration with reboiled sidestrippers. This is an unusual configuration but has benefits for high yields of adjacent products with large differential values.

Unit 2: The most heat-integrated crude unit the author has worked on. This is a high-temperature topping unit (atmospheric tower only).

Unit 3: A unit that has crude on the shell side of all its crude preheat exchangers. Only approximately 30% of crude preheat exchangers are configured this way, and not much data has been published on fouling for these cases.

Unit 4: A unit with two nearly identical parallel trains.

Table 2 summarizes the performance and configuration of these units.

Table 2 shows three outlet temperatures for each of the preheat trains. The first temperature (Actual) is the measured outlet temperature during the unit test run. The second temperature (TEMA) is the predicted outlet temperature if the exchangers were fouled at TEMA values. The third temperature (Clean) is the predicted outlet temperature if all the exchangers were clean.

Unit 3 had just reached one year of operation at the time of the test run. The other units had been operating for two years or longer. The exchangers were all in what would be considered normal cleaning cycles. Unit 3 is the only one that exceeded the heat recovery expected with TEMA levels of fouling. That may be due to the shorter run before data were gathered.

The four operating units included 105 heat exchangers in crude preheat service. Over 95% of the exchangers were operating with crude velocities >8 ft/sec (2.4 m/sec). This high velocity would generally be thought to contribute to lower fouling. In spite of this, three units showed performance that indicated fouling was higher than TEMA-recommended levels.

Actual fouling on exchangers in operating units depends upon how often the exchangers are cleaned. The greatest cost of fouling comes from lost capacity and possible lost yields. Energy costs are rarely a significant factor compared to the capacity and yield losses. This is true even when energy costs are high.

Table 2 Configuration and Performance of Operating Units Examined

	Unit 1	Unit 2	Unit 3	Unit 4
Crude In, °F (°C)	71 (22)	70 (21)	94 (34)	105 (40)
Crude Out, ° F (°C)				
Actual	424 (218)	572 (300)	442 (228)	477 (247)
TEMA	435 (224)	609 (320)	431 (222)	540 (282)
Clean	463 (239)	648 (342)	455 (235)	570 (299)
Exchanger FFs	ft²-°F-hr/BTU	ft²-°F-hr/BTU	ft²-°F-hr/BTU	ft²-°F-hr/BTU
Maximum	0.0132	0.0609	0.0100	0.123
Minimum	0.0014	0.0030	0.0000	0.000
Exchanger FFs	m²-°C/W	m²-°C/W	m²-°C/W	m²-°C/W
Maximum	0.0023	0.0107	0.0018	0.0217
Minium	0.0002	0.0005	0.0000	0.0000
Configuration	Atm-vacuum	Topping with preflash tower	Atm-vac with preflash drum	Atm-vacuum
Why included	Reboiled sidestrippers	Most integrated	Crude on shellside	Parallel trains
Services, total	8	18	9	17
Services, crude on shell	2	1	9	2
Services, crude on tube	6	17	0	15
Exchangers	10	41	12	42
Notes		1, 2	3	2

1: Clean performance limited by premature crude vaporization.
2: Clean performance limited by hot-outlet–cold-outlet pinch (HOCO).
3: No desalter in unit.

The units examined are all for North American locations with low energy costs. For these units, over 90% of the cost of fouling comes from lost capacity and yield. Dramatically increasing energy costs would only slightly increase the total cost of fouling. Thus conclusions from this analysis are applicable to high-energy-cost locations with only minor caveats.

Of course, cleaning frequency varies with how rapidly an exchanger fouls. One interesting observation was that there was no correlation between in-service fouling factors when exchangers are cleaned versus operating temperature, heat recovery fluid, or whether the crude was desalted or not. This is because exchangers that foul more rapidly get cleaned more often. For the most part, fouling is not allowed to proceed past some maximum value.

For the four plants shown, the tolerated maximum overall fouling factor was about the same even though the value of lost capacity and the cost of energy in each plant were vastly different. The value is also very close to the value shown by the fouling factors in the previous survey [1]. This surprising conclusion is worth a short explanation.

The largest cost of fouling is lost production, even when energy prices are high. A fouled exchanger still causes production loss when removed from service. Cleaning economics are driven by time out of service (production loss) versus improved production after cleaning. By comparison, cleaning costs and energy efficiency—while not ignored—tend to be less important. Capacity value (money/capacity) is assumed to be equal for the production loss and gain.

Since most heat trains were designed using a limited range of fouling values taken from TEMA recommendations, they encompass a limited range of likely duty shifts as individual exchangers foul or un-foul (get cleaned). The duty shifts strongly affect the economics of exchanger cleaning.

In spite of the differences between plants, the combination of design parameters, operating conditions, and economic factors all come together to effect an almost identical maximum fouling factor across preheat trains and in most units. To follow the logic:

- all (or very nearly all) exchanger designs used TEMA-influenced fouling factors;
- within a single preheat train, all (or very nearly all) exchanger designs use the same target approach temperature between the crude and preheat stream;
- all (or very nearly all) plants are driven by capacity losses or gains;
- incremental losses and gains have equal value per unit production; and
- all (or very nearly all) fouling trends are roughly linear over reasonable time periods. [1]

Cleaning occurs when gains > losses. The exact value of energy does not matter very much because it's small relative to other costs. The exact value of unit capacity does not matter very much because it's equal on each side of the capacity tradeoff. While it is true that changes in external forces (energy economics) can be expected to change the fouling value at cleaning, it would take enormous shifts to produce a significant change. Life cycle analysis of energy is outside the current scope but the basic observation is that it takes a lot of energy to make the steel for the exchanger and piping. As the value of energy goes up, the value of the recovered energy goes up, but so does the energy cost of building and maintaining the preheat train.

Many factors are unknown during design, but operating results and the above analysis show that the unknown factors balance out when looking at entire preheat trains. Based on extensive data, the average fouling value at cleaning is 0.015 ft^2-°F-hr/BTU (0.0026 m^2-°C/W). The author does not see any likely energy cost change in the near future that would significantly change this analysis, and therefore recommends using one-half of this value for the average as a guideline for design (0.0075 ft^2-°F-hr/BTU or 0.013 m^2-°C/W).

(Note that this value is not meant as a criterion for deciding to clean specific exchangers. In operating units, cleaning decisions should be made on a case-by-case basis for every exchanger.)

6. Fouling Factors Observed

Figure 2 shows the distribution of fouling factors for Units 1-4. Figure 3 shows the histogram of services versus fouling factor ranges. The observed fouling factors ranged from zero to 0.123 ft^2-°F-hr/BTU (0.0217 m^2-°C/W). The median fouling factor was 0.0074 ft^2-°F-hr/BTU (0.0013 m^2-°C/W). The mean fouling factor was 0.0155 ft^2-°F-hr/BTU (0.0027 m^2-°C/W). These median/mean values are very close to the previously identified values of 0.0075 ft^2-°F-hr/BTU (0.0013 m^2-°C/W) for regular services and 0.015-0.020 ft^2-°F-hr/BTU (0.0026-0.0035 m^2-°C/W) for critical services.

These high fouling factors for the exchangers in the survey are observed in spite of much higher crude velocities in the exchanger (>8 ft/sec, >2.43 m/sec) than are called for in standard design practices (2-4 ft/sec, 0.61-1.21 m/sec). Although higher velocity does reduce fouling in most crude oil services, the data here show average fouling factors at high velocities equal to the upper range of TEMA fouling factors for lower velocities.

No trend was seen for reduced fouling on the shell side versus the tube side. The only fouling-motivated reason for putting the crude on a specific side of a shell-and-tube heat exchanger is the relative ease of cleaning one side versus the other.

Fouling factors included in the study showed values of up to 0.123 ft^2-°F-hr/BTU (0.0217 m^2-°C/W), roughly 18 times the TEMA value for this type of exchanger and service. How severe can fouling be in operating units? High quality plant data from a unit not included in this or the previous study resulted in a calculated fouling factor of 0.145 ft^2-°F-hr/BTU (0.0255 m^2-°C/W)—the highest fouling factor ever personally verified by the author. This is 29 times the TEMA value for the given service in spite of nearly monthly cleaning of the exchanger. In this case, the exchanger is a filter protecting other exchangers in the preheat train. Cleaning it has practically no benefit for heat transfer.

7. Over-surface versus Fouling Factors

All the exchangers in Units 1-4 were designed using fouling factors. For each service, an estimate of the percentage over-surface was generated for each of three situations:

1. TEMA fouling
2. Recommended fouling value (.0075 ft^2-°F-hr/BTU, 0.0013 m^2-°C/W)
3. Actual fouling

The over-surface number is the percent extra area required that would let the exchanger reach the design operating duty if the fouling level matched the fouling factor for that situation. In addition, a duty-averaged percent over-surface for the entire preheat train was estimated for the same situations. An extremely fouled exchanger does little duty, so the duty-average tends to damp the contribution of extremely fouled exchangers.

Attempting to create a comparison between fouling factors and over-design depends upon choices in optimizing between exchanger layout, pressure drop, fluid velocity in the exchanger, and other factors. Generally, it is thought that there is no strict correlation for fouling factor versus over-surface specification that would give identical heat exchangers. The approach used here was to lengthen or shorten the exchanger, allow the pressure drop to vary, and keep the velocities the same. Table 3 summarizes these results.

Table 3 Duty Averaged Over-surface Requirements

	Unit 1	Unit 2	Unit 3	Unit 4
	Over-surface percent			
TEMA Fouling Factor	58	60	38	70
0.0075 ft^2-°F-hr/BTU (0.0013 m^2-°C/W)	111	71	41	82
Actual Fouling Factor	87	122	18	187

Figure 4 compares over-surface for specific exchangers versus the fouling factor; it shows a reasonable correlation between the two. This is most likely due to a consistent approach being used to determine the over-surface for all the exchangers, and contradicts a position held by many engineers who prefer to use fouling factors. In this regression, the coefficient of determination r^2=0.853, implying a fair amount of variance, but the fit remains within reasonable range for a heat-transfer correlation. Based on this, the rough equivalence between fouling factor and over-surface is:

Table 4 Fouling Factor and Over-surface

	Fouling Factor (Total)	Over-surface
	ft^2-°F-hr/BTU (m^2-°C/w)	Percent
Normal exchangers	0.0075 (0.0013)	72
Critical service	0.0150 (0.0026)	145

These values are meant specifically for the design approach of keeping velocity constant. The approach is the one favored by most proponents of over-surface.

It is important to note how large the over-surface areas are. Even the commonly used fouling factors of 0.004 to 0.006 ft^2-°F-hr/BTU (0.0007-0.0011 m^2-°C/W) translate into over-surfaces of 38% to 58%. The recommended values in Table 4 are much higher.

8. Implications of Demonstrated Fouling Levels

The demonstrated fouling levels far exceed the commonly used TEMA values. The higher values in TEMA are approximately equal to the average fouling levels shown. The implications are:

- Most preheat trains are under-surfaced for the heat recovery expected.
- Little evidence exists to support optimization that uses different fouling factors on each side of the heat exchanger.
- Overall preheat train performance would meet design criteria for most units if the higher TEMA values are used; TEMA temperature, velocity, and split between crude fouling versus preheat stream fouling is ignored; and a modest allowance is made for uncertainty.
- Most energy optimization studies underestimate the surface area required, leading to an underestimate of the incremental cost of energy recovery. Estimates of energy efficiency improvements in most crude units overestimate the economic optimum energy recovery point.

Table 5 summarizes the incremental duty recovered versus the incremental area required for designing to different fouling factors in the units. This is based on the fouling factor required to get the same performance as a clean unit. Table 5 supports three important observations:

- The units have a relatively low return of incremental duty versus incremental area. Again, this supports that the benefits for incremental heat recovery mainly come from capacity and, potentially, quality.
- The incremental energy recovery for these units falls within a small range.
- The calculated result for incremental duty recovered per incremental area added is nearly the same whether using the actual fouling factor or the TEMA fouling factor. With a properly optimized unit, the return on investment for using the proposed (larger) fouling values for design is roughly the same as or better than the return for using the TEMA values. This leads to a series of extremely interesting investment decisions beyond the current scope. The section of Future Work expands on this observation.

Table 5 Incremental Heat Recovery versus Surface Area

	Unit 1	Unit 2	Unit 3	Unit 4
If designed to	% incremental duty / % incremental area			
TEMA Fouling Factor	0.12	0.11	0.17	0.09
Actual Fouling Factor	0.11	0.11	0.20	0.11

In addition, the results in Table 5 raise questions regarding optimizing preheat trains in their entirety versus optimizing single exchangers. The analysis is based on only four units, so the support of the optimization questions is not as strong as the conclusions for overall fouling factors[b]. This would be another useful area for further work.

9. Future Work

The present analysis comes to conclusions that are in conflict with some commonly held positions on exchanger design and preheat train design. It has also opened some questions that merit further in-depth study. These include:

- What are the benefits of improving preheat train performance with chemical additives (reducing fouling rates), alternative exchangers types (reducing fouling rates), and cleaning techniques (reducing cost of cleaning)? Can these shift the overall average fouling level before cleaning?
- Are preheat trains being designed for the correct fouling levels? Based on the near-constant incremental benefit in surface area versus heat recovery regardless of fouling levels, is the industry using the correct economic analysis for preheat train energy recovery? While incremental energy recovery and the associated capacity improvements have a linear benefit, the incremental cost for surface area typically follows a less-than-linear scale. This suggests that if using the TEMA fouling recommendations can be justified, then it can be equally justified to use the higher fouling factors suggested here. Inversely, if the higher values here are not justified, then using the TEMA values is not justified either. The current analysis supports a decision to move away from TEMA recommended values—but from this work it is not clear, in which direction.
- At what point do pump costs and the hydraulics of extra surface area affect economic decisions on which fouling factors or over-surface to use?
- Is the industry using the best techniques to monitor exchangers and decide when to clean them? Is poor decision-making the cause of the observed fouling levels at cleaning?
- What can be learned from more work on comparing fouling factor versus over-surface approaches to preheat train design?

10. Conclusions

Analysis of fouling factors experienced in normally operating crude units was extended from previous work [1]. In the original study, fouling rates from 49 exchangers were used and total fouling was extrapolated by assuming a conventional 180-day to 250-day period between cleanings. The number of units involved in this dataset was unclear, but it was at least two or three. New data examined in the current analysis include fouling factors from 105 exchangers in four units. These

[b] Overall fouling factor conclusions are based on this work plus a reasonable sample of exchangers from additional plants.

units were all running in a roughly steady-state of cleaning-fouling cycles. Combining the data creates a total dataset of 154 exchangers from multiple units.

The average fouling factor between cleanings was nearly identical for the two sets of data. The data support a total fouling design value of 0.0075 ft^2-°F-hr/BTU (0.0013 m^2-°C/W). This is suitable for overall design of crude trains. Problem services, and exchangers in critical services, should use a value of approximately 0.0150-0.0200 ft^2-°F-hr/BTU (0.0026-0.035 m^2-°C/W). Critical services include those where low duty would lead to downstream problems on the non-crude side, and exchangers in the last one or two services in the crude preheat train.

Fouling factor data show little evidence to support lower design fouling values for any of the following scenarios: specific services, raw crude versus desalted crude, and temperature dependence of the fouling factor before cleaning.

It is clear that desalting and temperature affect fouling rates. The total fouling before cleaning depends upon both the fouling rate and the duration of fouling. The decision to clean an exchanger is based upon the expected gain from a cleaner exchanger versus the total cost of cleaning the exchanger (cleaning cost plus lost duty during the cleaning). The data show that the vast majority of exchanger cleaning occurs when the fouling factor reaches 0.0150 ft^2-°F-hr/BTU (0.0026 m^2-°C/W). This gives a time-averaged fouling factor of 0.0075 ft^2-°F-hr/BTU (0.0013 m^2-°C/W).

Desalting crude oil is critically important. Desalting increases unit capacity, reduces downstream corrosion problems, improves energy efficiency in the fired heater, and helps preheat train operation. For the preheat train, it is important because it reduces the fouling rate and lengthens the interval between exchanger cleaning. Desalter performance is vital in units that must shut down for exchanger cleaning.

Temperature does affect fouling, but there is no clear correlation. For some crude blends the worst fouling occurred when the crude was coldest. For other crude blends the worst fouling occurred when the crude was hottest or in the middle temperature range.

Joshi et al [5] claimed a significant correlation between shear stress (velocity) and fouling rates. The exchangers reviewed in this work neither support nor contradict that analysis. Nearly all the new data are from exchangers with high crude velocities (>8 ft/sec, 2.43 m/sec) so the data set has too narrow a range to support conclusions about velocity versus fouling rates.

Over-surface values were determined with a constant-velocity approach to exchanger optimization. Contrary to commonly held thinking, over-surface showed a reasonable correlation with fouling factors. Equivalent over-surface factors were determined as 72% for normal design and 145% for critical services. These over-surface values are much higher than often used.

Because they use too low a fouling factor or too low an over-surface, most energy optimization studies underestimate the surface area required. This means that the incremental cost used for anticipated energy recovery is too low. Most energy efficiency improvement projects in crude units overestimate the economic optimum energy recovery point.

The overestimates become especially costly with aggressively energy-integrated units. High-temperature energy recovery requires high metal temperatures. Higher temperatures require expensive materials of construction to avoid corrosion. If this costly surface does not capture the predicted energy, expensive mistakes can be made during design and revamp projects. Many revamp projects have failed to earn the return that justified them. Refiner projects are nearly always capital constrained. Using the correct basis makes sure that capital is used most efficiently. Understanding fouling is the key to correctly using capital in new crude units and crude unit revamps.

11. Figures

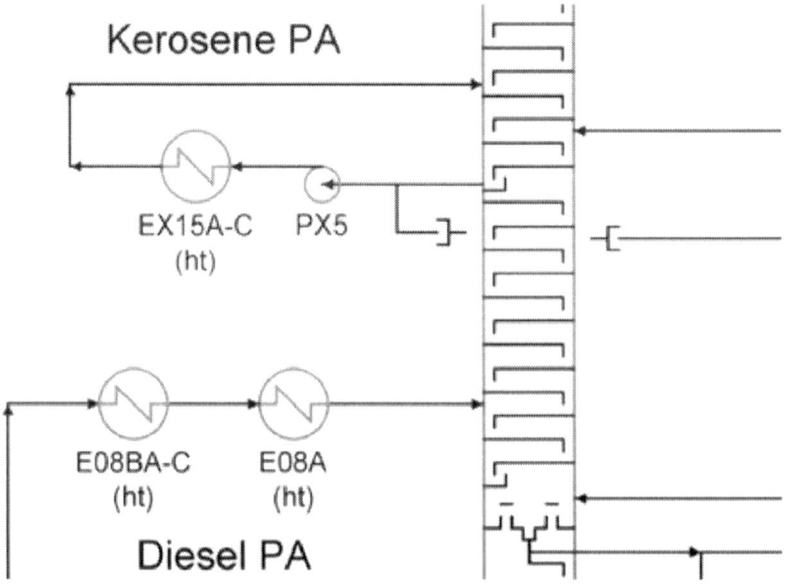

Figure 1 Preheat Train Exchangers Attached to Crude Distillation Tower

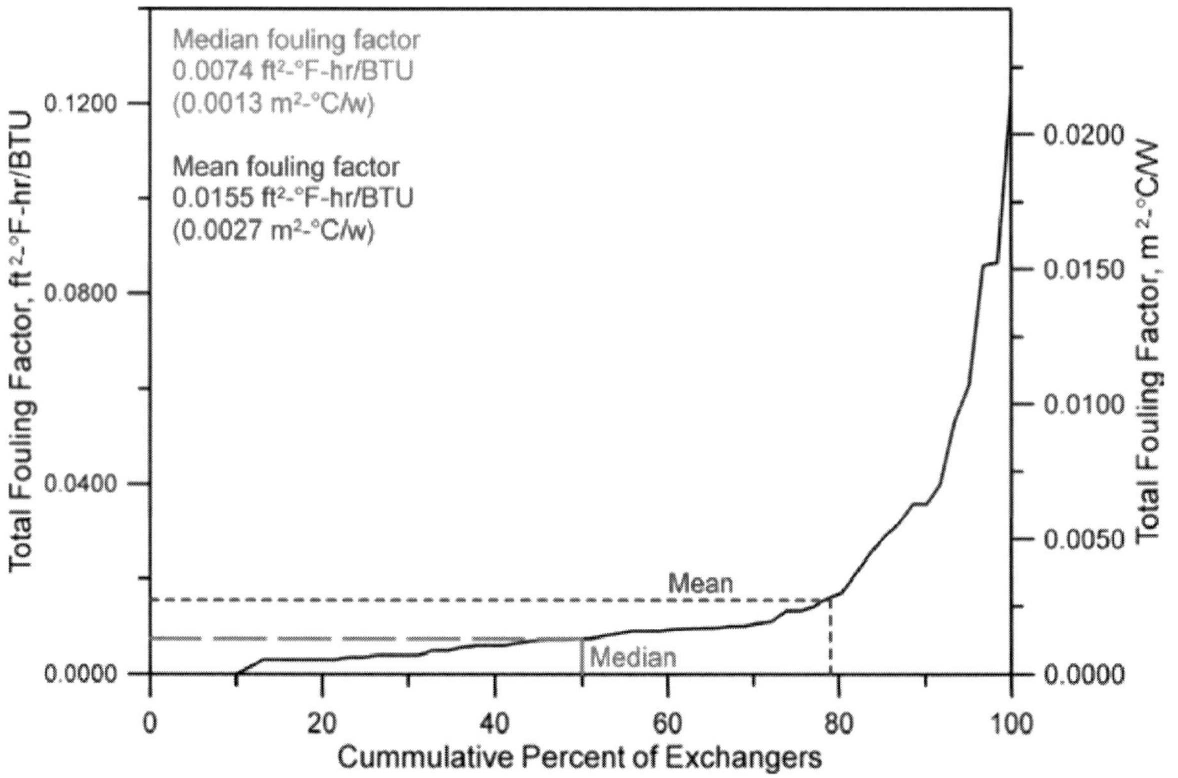

Figure 2 Fouling Factor Distribution

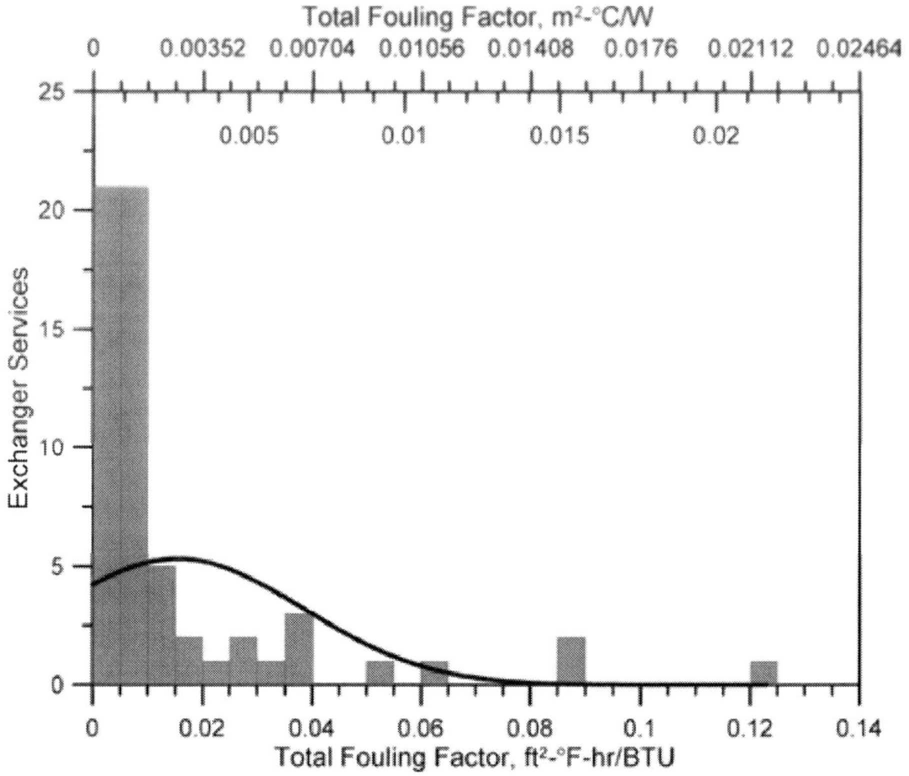

Figure 3 Exchanger Services versus Fouling Factor

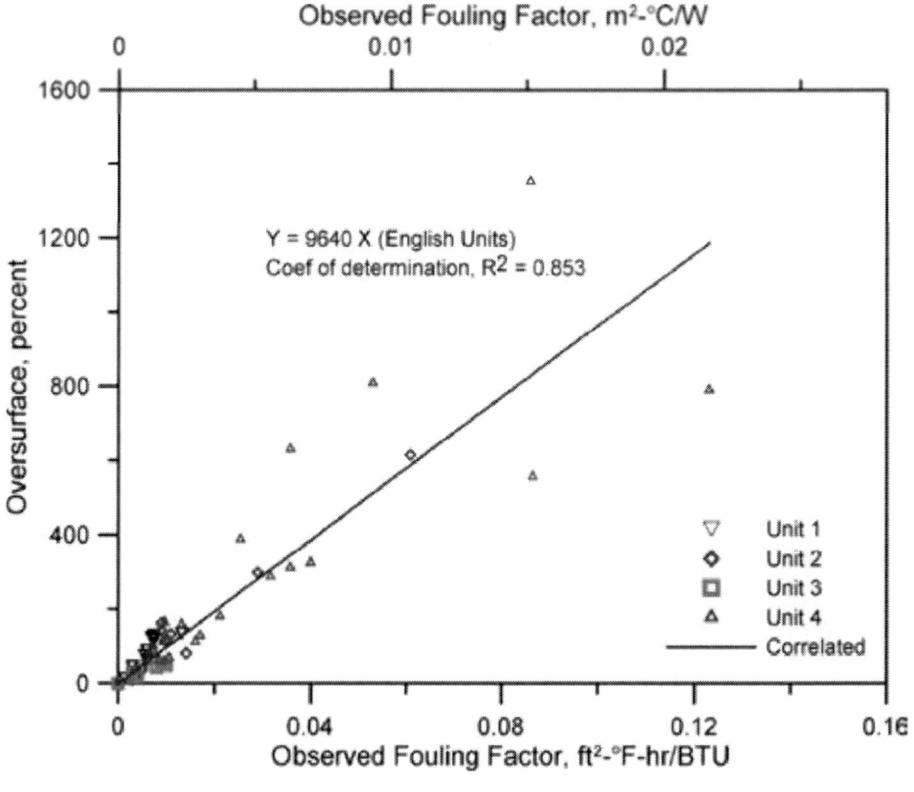

Figure 4 Over-surface versus Fouling Factor

12. References

[1] A. W. Sloley, "Crude Unit Fouling: What's Normal? Why's it Important?," presented at the AIChE Southwest Process Technology Conference, Galveston, TX, 2018.

[2] Anon., "Standards of the Tubular Exchanger Manufacturers Association, Ninth Edition," Tarrytown, NY, Standard, 2007.

[3] J. M. Chenoweth, "Final Report of the HTRI/TEMA Joint Committee to Review the Fouling Section of the TEMA Standards," *Heat Trans Eng*, vol. 11, no. 1, pp. 73–106, 1990.

[4] Various, "Proceedings Heat Exchanger and Cleaning Conferences (multiple)." Heat Exchanger Fouling and Cleaning, Various [Online]. Available: http://www.heatexchanger-fouling.com/proceedings.htm. [Accessed: 19-Feb-2019]

[5] H. M. Joshi, N. B. Shilpi, and A. Agarwal, "Relate Crude Oil Fouling Research to Field Fouling Observations," in *Proceedings of the International Conference on Heat Exchanger Fouling and Cleaning VIII*, Schladming, Austria, 2009 [Online]. Available: http://www.heatexchanger-fouling.com/papers/papers2009/2_Joshi_MS.pdf

550435 Use of Nonlinear and Machine Learning Techniques for Improved APC Modeling

Objectives

Model identification for advanced process control (APC) applications including model predictive control (MPC) and soft-sensor (or inferential) is one of the most time-consuming steps which requires dedicated plant tests and the skills of a highly experienced chemical engineering control engineer. The empirical models (such as first-order plus dead time, or FOPDT) need to be validated by judging the model attributes such as the gain value or sign, the dead time, and the shape of the step response based on prior process knowledge [1]. Plant tests must be repeated when the model identification does not match with the prior process knowledge.

To reduce the effort and cost of the model identification, this study proposes a simultaneous model fitting method which incorporates a priori information as constraints of the optimization problem. This approach can be applied to both linear empirical models and non-linear first principles models using the GEKKO optimization suite which specializes in dynamic optimization problems including nonlinear programming (NLP), differential algebraic equations (DAE), and artificial neural networks (ANN) [2].

Case studies and discussions

Two model identification examples are considered in this study. One is developing the inferential model for a soft-sensor in a distillation column. The other is a horizontal tank with a level control loop. Soft sensors play a critical role in estimating product quality (i.e. composition) in a real-time manner, especially in the chemical processing industry. They infer the product quality in the distillation column based on measurable column operating variables, such as tray temperature and pressure. The low frequency and time delay of on-line analyzers and lab sample results are not suitable for the real-time control applications. In the inferential fitting case, the industrial distillation column operation data is used to estimate the model parameters of the inferential model. The model is structured with three different blocks in series: the pressure compensated temperature (PCT) block, the FOPDT block, and the static nonlinearity block. For the nonlinearity block, a piecewise linear (PWL) function and an ANN are used and compared with a real-world result that uses a simple log transformation. Introducing the PWL and ANN as a nonlinear static gain block describes the nonlinearity of the process more accurately in an extended range of operation, which is difficult to achieve with a simple logarithm transformation.

In the tank example, instead of having a chain of model blocks, a nonlinear physics-based model is considered. The model is designed to accurately describe the nonlinear character of a horizontal tank with spherical caps. It has an inlet and an outlet flow with a PID level control loop configured on the outlet flow. The physics-based model consists of material balance equations combined with the tank geometry. The unknown parameters (such as the tank geometry) in the nonlinear model are estimated using the closed-loop operation data with the level controller.

Conclusions

In the inferential model fitting example, we investigate the hybrid approaches including piecewise linear and artificial neural net embedded into the simultaneous model fitting platform. The simultaneous model fitting by specifying constraints in a single optimization running reduces the extra effort to validate the model manually with a priori information and to manually sort out the appropriate segment of the historical operation data for each block. In the tank model example, the physics-based model is used to identify the unknown parameters using the available operation data. The identified physics-based model can then be used as a priori constraint in the simultaneous dynamic estimation problems in an empirical model fitting which is the most common type of advanced process control system. The benefit of this example is to improve the accuracy of the empirical linear model fitting and reduce the amount of time and effort by imposing the nonlinear constraint from the physics-based model as a priori information.

References

[1] Darby, M. L., & Nikolaou, M. (2012). MPC: Current practice and challenges. *Control Engineering Practice*, *20*(4), 328-342.

[2] Beal, Logan, et al. "GEKKO Optimization Suite." *Processes*6.8 (2018): 106.

COMPRESSOR SURGE MODELING AND CONTROL

Gregory K. McMillan, Emerson Automation Solutions, Chesterfield Missouri

Biography

Greg retired from Solutia Inc. as a Senior Fellow after 33 years specializing in improving loop performance. Greg contracted as a consultant in DeltaV R&D for 12 years and is presently a Principal Senior Software Engineer in Simulation R&D for Emerson Automation Solutions focusing on dynamic modeling using the Digital Twin. Greg is the author of 80 articles and papers and 20 books. Greg's most recent books are *Advances in Reactor Measurement and Control*, *Tuning and Control Loop Performance 4th Edition*, *Good Tuning: A Pocket Guide 4th Edition*, and *Process/Industrial Instruments and Controls Handbook 6th Edition*. Greg is an ISA Fellow and received the ISA "Kermit Fischer Environmental" Award for pH control in 1991, the Control Magazine "Engineer of the Year" Award for the Process Industry in 1994 and was inducted with Greg Shinskey into the "Process Automation Hall of Fame" in 2001. Greg was honored as one of InTech's 50 most influential innovators for advancing automation and control technologies and received the ISA Life Time Achievement Award in 2010.

Abstract

Axial and centrifugal compressor control is exceptionally challenging due to the extraordinary speed and severity of problems and the extreme consequences in terms of plant safety and performance. The fastest and most dangerous phenomenon is surge. An axial or centrifugal compressor in surge can reverse flow in 0.03 seconds going from a large positive flow to a large negative flow. Often, the negative flow is not measurable by flow meters leaving the actual situation to your imagination.

A high fidelity dynamic real time model has been developed and integrated into a Digital Twin to enable a greater understanding of the special requirements in terms of automation system speed of response and special algorithms. The model uses a momentum balance thoroughly researched and verified. The normally unseen compressor characteristic curve to the left of the surge point is used enabling the model to show the jumps in flow on the positive slopes of the characteristic curve. The unique shape and speed of surge flow oscillations is used in the Digital Twin to explore, develop, prototype, test, justify, deploy, commission and maintain the best surge control system. The Digital Twin also enables not only training of operators but education of process and automation engineers on the extreme dynamics and control system opportunities.

A simple but powerful technique is developed for detecting a potential surge and updating the surge curve. Control strategies are detailed that proactively prevent surge while maximizing compressor efficiency. Unnecessary surge valve openings are avoided and the surge controller setpoint is optimized to lie on the longitudinal axis of efficiency ellipses. Feedforward is used to deal with sudden disturbances from changes in production rates and shutdown of downstream users. Additionally, a method of optimizing the compressor discharge pressure based on user downstream valve positions is developed. All of these control system opportunities can be demonstrated with the actual control system connected to the model.

Compressor Surge Curve and Setpoint

A surge point is the point on the characteristic curve where the slope goes to zero. The characteristic curve is a plot of pressure rise versus suction flow for a given speed or inlet guide vane position. A volumetric suction flow is typically used because of advantages in terms of being more representative of the operating conditions for surge prevention. When a discharge flow is measured, temperature and pressure compensation are used to get it back to suction conditions. Operation at the point of a zero slope is extremely tenuous because a very slight deviation to the left results in positive feedback and a surge cycle. It is important that the operating point be kept sufficiently to the right on a part of the characteristic curve with a significant negative slope providing some self-regulation by negative feedback (e.g., increase in pressure causes an increase in flow and consequential decrease in pressure).

The surge curve is the intersection of surge points. For a centrifugal compressor, the surge curve is an exponential curve with an increasing slope for increases in flow demand. The ideal surge setpoint should be parallel to the surge curve. This provides a constant operating margin and aligns well with the long axis of the efficiency ellipses. The bias to the computation of the surge setpoint is minimized by making the measurements, surge valves, and PID controller fast enough to react for the worst case disturbance to prevent getting close to the surge curve. Coming off output limits and proportional relative to integral action is optimized.

For axial compressors, the surge curve bends over at high pressures. Also, the consequences of surge tend to be more disastrous because of the higher flows and the possibility of extreme acceleration. One very large compressor would hit a high speed shutdown in a fraction of a second potentially damaging the impeller that cost several millions dollars causing production losses of many more millions of dollars a day. A speed derivative was needed to detect the start of the acceleration and proactively stop the compressor.

Compressor Characteristic Curve and Surge Path

Often not shown is the compressor characteristic curve to the left of the surge point B where the slope becomes positive creating positive feedback causing the suction flow to jump to a negative flow point C on the compressor characteristic that has a negative slope. The operating point then walks along the negative slope portion of the characteristic curve due to negative feedback until it reaches point D the start of the positive slope and positive feedback at which point it jumps to point A, If the surge valve is not sufficiently open, the operating point walks back to point B starting the whole surge cycle over again. The precipitous drop in flow of the first jump can be as fast as 0.03 seconds. Each surge cycle tends to cause a loss in compressor efficiency that becomes noticeable after a total of 10 or more cycles. While successive cycles can be most damaging, the accumulation of cycles at different times is also a concern. The compressor manufacturer may have made tests to identify the surge point but may not have completely detailed the characteristic curves. We will see how surge points can be identified online and a first principle model can provide the characteristic curves to the left of the surge curve in sufficient detail to show the extreme oscillations during surge.

Flow and Pressure Oscillations during Surge

The flow oscillation is characterized by jumps from positive to negative flow and from positive to negative flow that correspond to the path on the positive slope of the characteristic curve creating positive feedback. The walk along the negative slope in the peaks and valleys take about 1 second leading to a total oscillation period of about 2 seconds for plant compressors. For laboratory and pilot plant compressors, the period is 20 times and 5 times faster. The fast periods for all compressors and incredibly fast jumps to very negative flows make recovery by just feedback control improbable.

The pressure oscillations have a period as fast as the flow oscillations but the jumps are slowed down and the amplitude attenuated by the discharge volume of the compressor. The best way of stopping the pressure oscillations is stopping the flow oscillations by an open loop backup that will hold the surge control valve open long enough for system to stabilize.

Compressor Surge Protection and Efficiency Maximization (Vent Valves)

A surge setpoint offset to the right of the surge curve on the compressor map is computed for a suction flow (surge) controller from the pressure rise across the compressor. The output of the PID surge controller goes to redundant vent control valves. The feedback controller is useful to prevent surge for slow approaches to the surge curve but needs help from control logic that is not operating in feedback (closed loop) mode to preemptively position and hold the surge valves open. Since this action is not closed loop and is meant to take over in potentially damaging situations, it is termed an open loop backup. When triggered, the open loop backup puts the PID surge controller in remote output putting the control valves in a position large enough to prevent surge.

The open loop backup must be able to detect surge as soon as possible. This is done by detecting the precipitous drop in flow that is the unmistakable indicator of the start of surge.

Since a surge cycle is extremely disruptive possibly causing the shutdown of downstream users and each surge cycle counts to a total that causes a noticeable loss in efficiency, it is extremely desirable to predict too close of an approach to the surge curve. This can also be done by the open loop backup. The key component for both recovery and prediction is a deadtime block to create a low noise reliable rate of change signal that can be updated as fast as the PID execution rate. For recovery from surge, a high rate of change triggers the backup that is the input minus the output of the deadtime block divided by the block deadtime that is typically just large enough to provide a good signal to noise ratio. For prevention of surge, a predicted flow one deadtime into the future that is too close to the surge curve triggers the open loop backup. The predicted flow is simply the input to a deadtime block plus the output with the block deadtime set equal to the total loop deadtime. Deadtime blocks can also be used to compute a pressure rate of change for a flow rate of change when operating to the right of the surge curve. A small pressure rate of change for a significant flow rate of change indicates an operating point close to the surge curve.

Turning on external-reset feedback (e.g., dynamic reset limit) enables the use of up and down rate limits on the setpoint the PID is manipulating that does not require retuning of the PID. External-reset feedback prevents the integral action from changing the PID output faster than what is happening in terms of the response to the PID output. This is called directional move suppression. Model Predictive Control greatly benefits from move suppression. By enabling a different move suppression depending on direction opens up many opportunities here and for many other applications. By putting up and down setpoint rate limits on the analog output block for the surge valves and fast readback of actual position, you can have a fast opening surge valve for surge recovery and prevention, and a slow closing valve for gradually returning to a more optimum operating position.

Directional move suppression is also useful for a valve position controller that seeks to lowers the compressor pressure to the point that keeps the furthest open downstream user valve at a good maximum throttle position. Directional move suppression by up and down setpoint rate limits on pressure controller will provide a slow approach to an optimum and a fast getaway for surge prevention. Not shown is a feedforward signal that would be added to the surge flow controller output to proactively open surge valves to a position that would balance a large decrease in flow by a downstream user. The feedforward signal would need to go through a signal characterizer that would compute the change in the X axis (change in valve signal) for a desired change in flow signal per the valves installed flow characteristic.

Compressor Surge Protection and Efficiency Maximization (Recirculation Valves)

The same kind of control strategies shown for vent valves can also be used for recirculation valves. There is a heat exchanger in the recirculation line to prevent heat buildup recirculating flow through the compressor. Recirculation valves are used for multi-stage compressors and compressors in series. It is desirable that stage or compressor have its own surge curve and control system. Decoupling signals can be used as necessary to prevent interaction. Compression rate dividing can be used to provide desired pressure rises for stages and compressors in series. If there are users of intermittent stage or compressor flows, feedforward signals would be applied to the respective stage or compressor surge controller.

Sources of Automation System Dynamics

It is critical that all the sources of delays and lags in a surge control loop be identified and minimized. The deadtime of the actual flow and pressure response (process response) is incredibly fast. The automation system is much slower and must be made as fast as possible to catch up with process response. The control valve often has the slowest 86% response time that is the summation of the pre-stroke deadtime and the rate limited exponential response (e.g., slewing rate with two lags). Signal filtering and transmitter damping must be minimized along with I/O scan times and filters and PID execution rates. Unsuspecting I/O filters are often the main limitation to the speed of response of a supposedly extra fast PID. The default values of these I/O filters may not be accordingly decreased to coincide with faster scan times offered. This mistake and many others are the result of not realizing all the potential contributors to the automation system deadtime that is the summation of all the delays plus equivalent deadtime from lags in series.

How to Make Valve Response Faster

Never replace a positioner with a booster despite age old rules that recommend this for fast loops. This can cause erratic dangerous movement (e.g., slamming shut) of a butterfly control valve from positive feedback due to high and low sensitivity of booster output and input, respectively.

A volume booster should be put on the positioner output with a bypass valve open just enough to prevent fast cycling by enabling the positioner to see some of the large volume of the actuator besides the small volume of the booster inlet port. The positioner should not have integral action and the backlash (deadband) and stiction (resolution) of the control valves must be minimized to less than 0.2% by the use of true throttling valves rather than tight shutoff on-off valve posing as throttling valves with deadband and resolution that can be order of magnitude or more worse (e.g., 2 to 10% deadband and resolution). The entire control valve assembly must be tested for small and large signal changes to verify the 86% response time is always faster than required (e.g., less than 1 second).

Valve Alerts

The most common mistake is selecting control valves based on tightest shutoff. These have the greatest seat or seal friction and consequently greatest stiction near the closed position. The next is maximizing capacity not realizing that any resolution limit or deadband is a percent of valve capacity and actual rangeability is lost preventing good control near the closed position. The other common mistake is seeking to minimize valve cost not getting the best most precise and responsive throttling valve.

There are many watch-outs. Actuators are often undersized based on thrust or torque requirement for loose packing or room temperature. The actuators should be sized to provide at least 150% of the thrust or torque required for maximum operating pressure and temperature. Resolution and deadband test should be done near the closed position where these effects are greatest (suppliers often choose 40% for testing valve response). The response time should be measured for small steps (e.g., 0.4%) and large steps (e.g., 40%) since actuator and positioner sensitivity and relay capacity come into play for small and large steps, respectively. On-off valves posing as control valves often have a shaft feedback that is not representative of ball or disk position due to friction in seal and packing and play in the actuator shaft to stem and from stem to actual ball or disk connections. A splined connection should be used for shaft to stem connection and the stem should be integrally cast with ball or disk (no connection). If this is not done, the smart positioner is being lied to by its position feedback giving diagnostics and readback that are not real leaving the user clueless.

Boosters should never be used instead of a positioner as previously detailed and integral action that makes the response to medium size steps look pretty, cause poor small and large valve response because the positioner gain needed is less than highest value to accommodate integral action. Also, integral action in the positioner will cause limit cycles from deadband when the surge controller is in automatic and cause limit cycles from stiction even when the surge controller is in manual.

Digital Twin Use and Improvements

The Digital Twin with a momentum balance to simulate the unique and very challenging requirements to prevent and recover from surge and automation system dynamics can be added to understand, design and verify compressor control systems maximize process safety, efficiency and capacity. The extreme consequences of less than the best control strategies and automation system components can be detailed and corrections can be made by rapid exploration and testing. The only limit is your imagination. Online metrics can show the benefits in terms of less loss of compressor efficiency and better process performance.

Digital Twin's Key New Blocks

Blocks have added to the Digital Twin to test innovations and improvements in strategies, dynamics and PID tuning. The Compressor W/Surge block uses a momentum balance to provide the dynamics of surge. The block can be put in series for multi-stage compressors or compressors in series. A Backlash-Stiction block can be put in series with each other to simulate what is happening in the actuator, shaft to stem connection, and stem to ball or disk connection. Blocks to include a variable deadtime, slewing rate and response lags can be inserted. The Future Value block can provide a timely low noise rate of change and future value typically a factor of the total loop deadtime into the future from current value. This is extremely useful for surge curve and actual surge identification.

Details on Control System Implementation and Best Practices

The principles and knowledge are listed as implementation details and best practices.

Conclusion

This is just a brief insight into the incredible deep understanding gained and great advances possible by compressor modeling and control.

References

1. McMillan, G.K., Stuart, C., and Hildebrand, T., "Compressor surge control: Deeper understanding, simulation can eliminate instabilities", *Control*, Oct 2018 https://www.controlglobal.com/articles/2018/compressor-surge-control-deeper-understanding-simulation-can-eliminate-instabilities/?stage=Live
2. McMillan, G.K. and Berutti, M.E., "Virtual plant virtuosity", *Control*, April 2017 https://www.controlglobal.com/articles/2017/virtual-plant-virtuosity/?stage=Live
3. McMillan, G.K. and Wu, P., "How to specify valves and positioners that don't compromise control", *Control*, May 2016 https://www.controlglobal.com/articles/2016/how-to-specify-valves-and-positioners-that-dont-compromise-control/?stage=Live
4. McMillan, G.K., *Centrifugal and Axial Compressor Control*, Momentum Press, 2010 http://www.momentumpress.net/books/centrifugal-and-axial-compressor-control

549081 A Model-Centric Framework for the Smart Manufacturing of Polymers

Plastic and polymeric materials have an enormous impact on a variety of industries and our daily lives. The synthesis of polymers occurs by reacting small molecules referred as monomers together to form, under certain conditions, long molecules referred as polymers. During polymerizations, it is mandatory to assure the uniformity between batches, high-quality of products, energetic efficiency, minimum impact to the environment, reliability of the process, and labor safety. It remains as a major challenge the establishment of operating, monitoring, and control conditions capable to achieve all mentioned goals together.

In this work, various model-centric strategies towards the smart manufacturing of polymers were designed, implemented, and tested using experimental data from two polymerization systems. The main aims included the combination of different model-centric strategies which permit to obtain operational conditions such that production goals are achieved. After validating the correctness and reliability of the fundamental model, different optimal policies are formulated, advance control elements are developed, and nonlinear online estimators are designed for reaching the joint goal of signal processing and full polymer characterization.

The first system studied is the synthesis of polyacrylamide in aqueous solution using potassium persulfate as initiator in semi-batch fashion. In this system, the proposed framework integrates nonlinear modelling, dynamic optimization, advanced control, and nonlinear state estimation towards producing polymers with target properties. The main objectives include the attainment of desired polymer characteristics through feedback control while a complete polymer motoring is possible during the reaction. Results, in terms of the dynamic set point achievement, are satisfactory for the closed-loop system. An improvement of 42% in average is foreseen when using a hybrid discrete-time extended Kalman filter (*h*-DEKF) as state observer. In addition, other strategies for state estimation are proposed including different structures of the geometric observer (GO) from which the 4-state GO with passivation shows the best performance overall. The 4-state GO with passivation achieves an improvement of 52% in the estimation of the molecular weight distribution when compared with the *h*-DEKF.

The second system is the copolymerization of ethylene with 1,9-decadiene using dimethylsilyl (N-tert-butylamido) (tetramethylcyclopentadienyl) titanium dichloride (CGC)/MAO as catalyst in semi-batch fashion. For this system, a data-driven strategy for online estimation of important kinetic parameters was assessed. A global sensitivity analysis was performed for selecting the most representative kinetic parameters of the system. The retrospective cost model refinement (RCMR) algorithm is adapted to the problem and implemented for online parameter estimation. After verifying the stability and robustness of the method, real experimental data permitted to validate its performance. Results show that the estimation of the kinetic parameters of interest is feasible, and converges close to theoretical values without requiring previous knowledge of their magnitudes.

550628 Synthesis of Reactive Separation Systems Via Building Blocks

Process intensification (PI) has become an important tool for novel process designs featuring drastic improvements in process performance. While miniaturization, use of alternative energy sources and discovery of multifunctional materials are examples for activities leading to intensification, integration of several operations in a single equipment have been one of the most successful PI strategies in terms of their industrial acceptance [1-2]. Specifically, reactive distillation (RD) units, which combine reaction and distillation operations in a single column, have seen a marked increase in their industrial adoption since 1980s [2]. Wider utilization of this technology requires systematic design procedures suggesting optimal intensified designs at the early design stage. Several methods have been suggested in the past for the synthesis of reactive distillation operations [3-4]. In this work, we use building block-based representation [5-8] to provide a systematic procedure for the synthesis and optimization of reactive separation processes. This novel representation can suggest non-intuitive intensified configurations with its unique outlook towards process design. The representation is based on collection of blocks positioned on a two-dimensional grid which can be used to represent several physical and chemical phenomena. Each building block has several physical attributes, i.e. temperature, pressure, phase, composition, and can stand for a unit use of a material, e.g. catalyst. While reaction operations can be represented within a single block, separation related phenomena require at least two blocks with different phases. Vapor-liquid equilibrium, for instance, requires one vapor and one liquid block sharing a common boundary through which phase transition can take place. When reaction operation is assigned into one of these neighboring blocks, a reactive vapor-liquid contact stage can be obtained. Collection of these building blocks in a grid formation yields building block superstructure which is modeled with a Mixed Integer Non-linear Programming (MINLP) model. Continuous variables are used to quantify the material and energy flow between the blocks and integer variables are utilized to assign different phenomena at different positions on the grid. This results in a generic representation and synthesis method for reactive separation processes. We will demonstrate the applicability of the proposed method with several example problems including ideal and non-ideal reactive systems and show that cost-optimal intensified designs can be realized through building-block based approach.

[1] Stankiewicz, A. I., Moulijn, J. A. (2000). Process Intensification: Transforming Chemical Engineering. Chemical Engineering Progress 1, 22–34.

[2] Tian, Y., Demirel, S. E., Hasan, M. M. F., Pistikopoulos, E. N. (2018). An Overview of Process Systems Engineering Approaches for Process Intensification: State of the Art. Chemical Engineering and Processing: Process Intensification, 133, 160-210.

[3] Ismail, S.R., Pistikopoulos, E.N. and Papalexandri, K.P. (1999). Synthesis of reactive and combined reactor/separation systems utilizing a mass/heat exchange transfer module. *Chemical engineering science*, *54*(13-14), 2721-2729.

[4] Okasinski, M.J. and Doherty, M.F. (1998). Design method for kinetically controlled, staged reactive distillation columns. *Industrial & Engineering Chemistry Research*, *37*(7), 2821-2834.

[5] Demirel, S. E., Li, J., and Hasan, M.M.F. (2017). Systematic Process Intensification using Building Blocks, *Computers and Chemical Engineering*, 105, 2-38.

[6] Li, J., Demirel, S.E. and Hasan, M.M.F. (2018). Process synthesis using block superstructure with automated flowsheet generation and optimization. *AIChE Journal*, 64(8), 3082-3100.

[7] Li J., Demirel S.E., Hasan M.M.F. (2018). Process Integration using Block Superstructure. *Industrial & Engineering Chemistry Research*, 57, 4377–4398.

[8] Li J., Demirel S.E., Hasan M.M.F. (2018). Fuel Gas Network Synthesis Using Block Superstructure. *Processes*. 2018, 6(3), 23.

550417 A Comparative Study of Residence Time Distribution and Cooling Crystallization in a Continuous Dynamic/Oscillatory Baffle Crystallizer Versus a Stirred Tank

Crystallization has become a predominant technique in particle design technology and it is widely applied in the pharmaceutical industry. Traditionally crystallization has been carried out in batch mode where operating conditions vary with time, often resulting in inconsistent products.[1] Continuous crystallization, on the other hand, has been identified to improve reproducibility, reduce operating cost and simplify scale-up.[2] A novel commercial dynamic baffle crystallizer(DBC) operated as a mixed-suspension-mixed-product-removal(MSMPR) system has been studied. It consists of a reactor tank and dynamic 'donut' shaped baffles to provide oscillatory flow resulting in more uniform mixing.[3] Oscillatory systems have been well studied for batch reactions and synthesis but few studies have been carried out to study the DBC for continuous two-phase operations such as crystallization.

In the work proposed here, liquid and solid residence time distributions were studied systematically in the DBC in comparison with a traditional continuous stirred tank crystallizer(STC). In-situ process analytical technology(PAT) tools such as UV-vis, Infrared(IR), and focused beam reflectance measurement(FBRM) were used to measure system response to step or pulse changes in order to construct RTDs. It is observed that the uniformity of both liquid and solid RTDs was significantly improved in the DBC than the STC indicating potential improvement in size uniformity during crystallization processes. A preliminary continuous cooling crystallization of paracetamol was carried out for two different oscillating conditions in the DBC as well as the STC at the corresponding rotational speeds yielding (approximately) equal power density. Online PAT tools were used to monitor the processes in addition to sampling for offline characterization at every residence time (45min). Nucleation event took place earlier in the DBC than the STC possibly due to the improved mixing mechanism. At early stages of operation, DBC produced crystalline particles of comparable or improved size uniformity. At later stages of operation (>7 residence times), particle sedimentation or severe particle breakage occurred in the STC causing process deviation from steady state while such events did not take place in the DBC and steady state was maintained. In the RTD studies as well as the preliminary crystallization study, the DBC showed improved system uniformity and consistency demonstrating potentials for more in-depth crystallization studies in the DBC.

1. Nagy ZK, Aamir E. Systematic design of supersaturation controlled crystallization processes for shaping the crystal size distribution using an analytical estimator. *Chem Eng Sci*. 2012;84:656-670. doi:10.1016/j.ces.2012.08.048.
2. Alvarez AJ, Singh A, Myerson AS. Crystallization of Cyclosporine in a Multistage Continuous MSMPR Crystallizer. *Cryst Growth Des*. 2011;11(10):4392-4400. doi:10.1021/cg200546g.
3. Hewgill MR, Mackley MR, Pandit AB, Pannu SS. Enhancement of gas-liquid mass transfer using oscillatory flow in a baffled tube. *Chem Eng Sci*. 1993;48(4):799-809. doi:10.1016/0009-2509(93)80145-G.

546606 Design of Vapor Recompressed Dividing-Wall Columns Processing a Light Component Dominant Mixture

When a dividing-wall column (DWC) processes a mixture with a dominant composition of the light component, there exists not only a relatively small temperature elevation from the overhead to the stripping section of its prefractionator but also a large amount of latent heat available in the overhead vapor. These provide opportunities for applying heat pump to the DWC and lead potentially to great improvement in system performance. In this paper, a DWC processing an N-pentane/N-hexane/N-heptane ternary mixture is chosen to examine the characteristics and effect of the applications of heat pumps. We use Stage-Enthalpy Deficit (S-H) profile and Temperature-Enthalpy Deficit (T-H) profile to guide the combinations of heat pumps with the DWC while taking the minimization of total annual cost (TAC) as an objective function. The obtained result shows that multiple heat pumps might be employed to use the compressed overhead vapor to heat the feed, the stripping section of the prefractionator, and the section above the intermediate product of the main column. The first heat pump is employed to preheat the feed, thereby allowing the adoption of feed splitting to strengthen the contact between vapor and liquid phase in the prefractionator. The second heat pump is installed to heat the stripping section of the prefractionator and the third heat pump is installed to heat the separating section above the intermediate product of the main column. The TAC decreases by 25.09% as compared with the DWC. The total utility consumption (TUC) decreases by 62.55% as compared with the DWC. These striking results highlight the great potential of applying heat pump to facilitate the DWC. Current work is underway to simplify the search process in process synthesis and design.

550403 Cost Estimating Benchmarks for Brownfield Automation Capital Projects. Determining What Your Project Really Costs

Early in the financial feasibility stage some idea of total installed costs are needed to proceed. How project costs are affected by scope decisions is key information for informed decisions. This means some knowledge of the scope of work and associated costs for modernization projects is required. However engineering and planning of automation modernization projects is not a well known practice. In most brownfield plants a modernization project of any significant size may have rarely been done recently at your site. This presentation addresses historical installed costs associated with modernization, ie 'rules of thumbs' for automation projects. Topics addressed include what make brownfield projects different than greenfield, the associated activities needed for brownfield, how installed costs breakout among equipment, engineering and installation, the effects of project scope and what factors you can manipulate to control cost.

549920 The Journey Towards 2030, Are We There Yet? An Impact of 2030 Vision and 2020 Transformation Program on Energy, Infrastructure and Other Projects in the Region

Saudi Arabia and other Gulf countries announced their vision for 2030 trying to have better future for next generations and prepare for the after oil ERA.However, some countries are easing the transformation to avoid falling in recession. How will this effect announced and new projects? In this paper we will see the effect of this vision and transformation program and other programs on the projects in the region during the past period and see how this effect the oil would and gas business and projects in the region? Will this mean more projects for the energy sector and petrochemicals or are countries moving away from there? Will oil prices and budget deficits postpone or otherwise accelerate this transformation? If yes, then for which projects? Will we see more fuel expansion projects? Renewables? Solar? Will entertainment, infrastructure and housing projects have the highest priority?

551252 Oil-to-Chemicals Pathways Renewed By Separations and Catalysis

Main products from crude oil are motor fuels. The growth of motor fuel products is projected to decelerate under the pressure of hybrid and electric-drive vehicles, biofuels and gas-based fuels such as LNG, CNG, LPG and methanol. At the same time, the global consumption of chemicals continues to grow. Oil-To-Chemicals pathways and corresponding integration of a refinery with petrochemical units has garnered attention with refiners, majors and national oil companies possessing of large reserves of crude oil. This presentation reviews existing, former and possible future Oil-To-Chemical pathways. It examines the impact of improved separation methods, paraffin conversion to chemically active species and olefin derivatization catalyst development on the viability of these paths. Separation and catalyst development needs for possible future Oil-To-Chemical pathways are identified.

550132 Aligning Business Drivers to Create Win-Win Water Reuse Partnerships between Refineries and Municipalities

Background

Historically, water has been a low-cost utility for many refiners – one that required little consideration and optimization as to the source. However, as water stress in industrial facilities has become more prevalent in recent decades, refiners have been forced to more closely examine their water usage. Municipal wastewater effluent is an intriguing water source for refineries trying to reduce their reliance on potable water sources. It has been successfully used for makeup water in numerous facilities, allowing these refineries to respond to water stress and reduce their water costs. However, there exists a natural conflict between industrial users and municipalities when it comes to water reclamation. While reuse of municipal wastewater offers a refinery a lower cost source to reduce their expenditures on water, the refinery's gain can become the municipality's loss as decreased revenue is realized from the sale of potable water. As a result, many discussions regarding water reuse begin with great intentions, but ultimately never result in an agreement due to failure to address key business drivers of all stakeholders.

Alignment of Business Drivers

The inherent conflict of interests plays a significant role in the feasibility of municipal effluent reuse, and a successful reclamation program is one that alleviates this conflict to provide a sustainable solution that is a win-win for both parties. While the business interests of refineries and municipalities may diverge at times, they find common ground on the complexity of the issues they face. Environmental and human interests, water scarcity and conservation, the infrastructure present, and reliability drive the decisions of municipalities and refineries, and ultimately those interests must balance with revenue and capital expenditures required for additional infrastructure. Although the large number of factors spurring wastewater decisions certainly can complicate the path toward a successful water reclamation program, they also provide ample opportunities to find similarities in the issues that both parties face and reach an agreement that allows each to come out ahead. While negotiating these win-win reuse agreements can be a difficult task, the ability to understand and appeal to the business drivers of the other party can go a long way toward reaching a deal.

When it comes to the varied business drivers of refineries and municipalities, there are certainly factors that can push them into alignment. Water stress is one of the most common, with drought, population growth, and degradation in the quality of existing water sources all being observed in various parts of the county. In these instances, refiners and municipalities have a clear reason to cooperate in conserving their local resources and preserving their business. These concerns are not going away any time soon. Water stress is expected to increase across the United States, and so will the need for reliable alternative water sources. It can be expected that the number of water reuse programs between refiners and municipalities will increase in conjunction. There is a need to secure water resources to ensure the long-term health of the business, which ultimately boils down to the same idea for both municipalities and refiners – a sustainable and consistent water source is vital for profit generation.

Municipalities and refineries also share a need for long-term planning of infrastructure upgrades. Water reuse solutions require capital expenditure for infrastructure, and both parties must reach an agreement on the responsibilities for expenditures. Negotiating capital expenditures associated with water reclamation can be difficult due to varied expectations in return on investment between industry and municipalities, as well as potentially complex financing options that may be available. A long-term partnership via a water reuse program can provide stability to allay these concerns, while also promoting greater coordination to allow for capital efficient solutions and optimization of treatment. With greater assurance that both parties will see a return on their investment via a long-term agreement, municipalities and refineries can feel more secure in the capital expenditures they make.

Case Studies

A number of refineries in water stressed regions of the country have entered into water reuse partnerships with local municipalities to great success. One midwestern refinery used municipal effluent to reduce its water draws from a nearby aquifer, which also supplied raw water to the city's potable water treatment plant and was being dewatered at an alarming rate. As a result of the project, the refinery was able to reduce its spending on raw water and secure a water source for years to come, and the municipality was able to preserve the health of the aquifer and protect the source of potable water for its inhabitants. Other municipalities in the water-stressed west have responded to booming populations and drought conditions by implementing their own reuse programs with local refiners. One bay area municipal water district has partnered with a nearby refinery to provide municipal wastewater effluent to the refinery for cooling tower makeup and boiler feed water at no volumetric charge (Baldwin and Towey, 2015). In exchange, the refinery reimburses the municipality for all capital, operating, and maintenance costs. With the institution of the reuse program, the refinery avoids drought surcharges and has a more cost effective raw water source while demonstrating their commitment to sustainability, and the municipality frees up potable water sources for human consumption while receiving financial backing to complete projects that improve their ability to protect their water sources for the future.

Conclusion

The reuse of municipal wastewater effluent provides a sustainable method of conserving water resources and has been executed with success in numerous instances around the country. While arriving at a reuse agreement that is a win-win for both parties can be difficult, there are numerous instances in which municipalities and refiners can find shared business objectives. By appealing to these shared objectives and understanding the drivers of the other party, refiners and municipalities have reached agreements that preserve water supplies, future business, and communities.

References

Baldwin, L. S.; Towey, A. Downstream Oil and Gas Water Reuse. Presented for WateReuse Webcast Series [Online], June 24, 2015; WateReuse. https://watereuse.org/wp-content/uploads/2015/10/DownstreamOill-and-Gas_webcast.pdf (Accessed Nov 29, 2017).

551245 Impact of Nutrients on Water Environment: Control Technologies and Policy

Effluent streams of domestic origin contain nitrogen (N) and phosphorus (P) that are essential nutrients for the growth and metabolism of microbial population in biological wastewater treatment process. Effluent streams for certain chemical industries contain high concentrations of ammonia. These include fertilizer production plants and petroleum refineries. While parts of these nutrients are utilized in cell growth, the remainder is discharged with the treated effluent. If this is discharged to a confined body of water, e.g. lake, estuary, or a bay it fosters undesirable aquatic growth of algae, hyacinths, etc., which adversely impacts the quality and aesthetic of the receiving body of water. Increasing number of states in the US are implementing stringent effluent discharge regulations for ammonia and total nitrogen.

This paper will present case studies on the steps taken and innovative technologies implemented to control nitrogen from wastewater treatment plants in the US. These will be supplemented by a discussion of technologies for reuse and resource recovery from treated effluent.

551491 Investigation of Hydrocarbon Utilization By *Debaryomyces Hansenii* (LAF-3 10 u) in Desalter Effluent

The bio-remediation of desalter effluent in the crude oil refinery and its reuse in the petroleum refinery is a practical option. Desalter effluent has a high concentration of salt which is extracted from crude oil in the petroleum refinery. The effluent contains BTXs components and other hydrocarbon compounds which are non-aqueous phase liquid and not immiscible in water. The salinity of effluent, immiscibility of BTXs and other hydrocarbons results in subsequent issues in downstream treatment of wastewater. On the other hand, BTXs are serious carcinogenic and toxic effects, which are affecting the respiratory system. In this study, *Debaryomyces hansenii* , a species of halophilic yeast was selected for its tolerant in the saline effluent. In this study, nonionic surfactant was added to investigate BTXs and dodecane to be exposed to the yeast *Debaryomyces hansenii*. The biodegradability of BTXs and dodecane in high concentration range in comparison to other studies for BTXs removal as a carbon source for *D.hansenii* in Desalter effluent was investigated. The kinetics of BTX degradation by *Debaryomyces hansenii* are presented.

550267 Optimization of Alkaline Peroxide Pretreated Rice Husks Using Response Surface Methodology

Cruse oil released into the ocean can lead to both environmental pollution of the water system and economics lost for the water purification. Rice husk is a desirable resource for extensive utilizations, such as oil sorption. However, with the high concentrations of lignin and silica, alkaline peroxide treatments have been applied to rice husks. The delignification of rice husk under alkaline peroxide condition is developed for the removal of lignin and silica. The ash content in rice husk is also considered as one of the major obstacles to biofuel generation.. As the main component of ash in rice husk, silica has been known to exist as a hard shell covering and protecting the whole structure of rice husk, which can lead to process operational issues. Although a number of sporadic studies regarding the alkaline hydrogen peroxide (AHP) treatments on lignocellulosic materials in recent years, a parametric research approach applying Response Surface Methodology (RSM) has not been applied to study the delignification performance and the leaching of silica ash from rice husks. For instance previous study has used RSM to study the removal of lignin from vineyard pruning residues by AHP treatment, but because of the difference of silica concentrations in vineyard pruning residues, the silica content was unfortunately not considered as a response factor. As a statistical technique for modeling, RSM is usually used to analyze the performances of designated independent variables on selected response variables. By applying RSM designs, statistically valid results can be obtained by only conducting a limited number of experiments. With such effectiveness and popularity, RSM has been adopted widely in different fields.

In the present work, rice husks were pretreated by NaOH combined with H_2O_2. High Performance Liquid Chromatography (HPLC) analysis was undertaken as the basis to investigate the reaction. RSM based on Box-Behnken statistical experiment design was applied to optimize operating conditions, which are suitable for maximum lignin and silica removal while maintaining most cellulosic components in the solid residues. Hence, as a result such modified rice husk residues will turn out to be more suitable for the utilization of energy production. Three independent variables of NaOH concentration, H_2O_2concentration and reaction temperature were studied by using Box-Behnken design. The optimized conditions are discussed according to two different criteria. The first one focused on the productivity and the optimized conditions are 8% NaOH, 1% $H_2O_2$and 20 °C. The second one emphasized the cost effectiveness and the optimized conditions are 5.29% NaOH, 1% $H_2O_2$and 20 °C. The validation experiments additionally confirm the reliability of the predicted model.

When emphasizing the productivity, the optimized conditions are 8% NaOH, 1% $H_2O_2$and 20 °C. When cost effectiveness is included, the optimized conditions are 5.29% NaOH, 1% $H_2O_2$and 20 °C. The 50.89% actual solid yield under criteria (a) has confirmed the successful treatment when considering the initial concentration of different components in rice husk (28.21% lignin, 16.5% ash and 5.01% extractives). The effectiveness of the designed experiment suggested rice husks can be considered as a possible sorbent candidate to relieve the water pollution issue cause by oil spills.

551345 Energy and Exergy Analysis of a Single Evacuated Tube Heat Pipe Solar Collector Applicable for the Solar Powered Desalination

In this work, we investigated single evacuated tube heat pipe solar collector (ETHPSC) for the efficient utilization of solar energy for thermal pumping purpose by using novel Encontech heat engine towards reverse osmosis desalination. The effect of inlet mass flowrate , supplied temperature , solar intensity , ratio of condenser to evaporator section area , heat transfer coefficient on the efficiency and collector temperatures is studied in detail by using developed heat transfer models. The simulation results shows that, each has optimum based on exergetic performance. Likewise, simulation results revealed that the lowered h_{Vaccum} is found to be more promising to improve both energetic and exergetic efficiency and temperature gain wherein the maximum energetic efficiency of 52.52 % was found for = 1 . Furthermore, simulated results specifies that for of 0.3 and of 50 demonstrates more energetic and exergetic (=31.4 %, =2.6 %) performances in compare to of 100 (=11.8 % =2.15 %). Furthermore both efficiency increases approximately till equal to 1 and then reduces with 1. In overall, simulation records showed that the energetic performances improved with solar intensity and mass flowrate at lower inlet temperature and heat transfer coefficient. To achieve the better performance, these results are more useful towards appropriate selection of with consideration of optimized inlet temperature and heat transfer coefficient.

550549 Monitoring Process Degradation through Operating Regime Based Process Monitoring

Process monitoring is an essential component of many industrial processes, and is required in order to maintain process safety, and minimize operational costs [1]. Process degradation through equipment deterioration can lead to a significant increase in operational costs, and thereby result in high economic losses.

Monitoring equipment deterioration can be a challenge as these parameters cannot be directly measured. Fortunately, state estimation methods are available, which can be utilized in order to estimate the values of these process parameters, and additionally other state variables, including those that cannot be easily measured using sensors. The family of Kalman filter estimation methods, including the Extended Kalman Filters (EKF) and Unscented Kalman Filters (UKF), have been utilized by many in order to carry out effective state and parameter estimation for different applications [2]. Many industrial processes, and pieces of equipment are known to follow nonlinear models. Recently, state estimation through particle filters has also received a reasonable amount of attention.

This work involves monitoring of a process using state estimation methods in order to identify the economic impact of equipment deterioration over time. Application of the algorithm will be demonstrated through a nonlinear example, e.g., a heat exchanger that experiences increasing levels of fouling over time. The presence of fouling causes the overall heat transfer coefficient to decrease, making the heat exchanger less efficient. This in turn forces the heat exchanger to utilize more cooling utility in order to maintain the same set points, thereby increasing operational costs. The example will estimate the parameter, i.e., overall heat transfer coefficient, and the states using different state estimation methods [3]. Additionally, dynamic data will be obtained by simulating the nonlinear heat exchanger in ASPEN Plus Dynamics, in order to enhance of the validity of the proposed algorithm.

References

[1] M.Z. Sheriff, F. Harrou, M. Nounou, Univariate process monitoring using multiscale Shewhart charts, in: 2014 Int. Conf. Control. Decis. Inf. Technol., IEEE, Metz, France, 2014: pp. 435–440.

[2] M.M. Mansouri, H.N. Nounou, M.N. Nounou, Nonlinear control and estimation in induction machine using state estimation techniques, Syst. Sci. Control Eng. 2 (2014) 642–654. doi:http://dx.doi.org/10.1080/21642583.2014.956842.

[3] M.S. Peters, K.D. Timmerhaus, Plant design and economics for chemical engineers, 4th ed., McGraw-Hill, 1991.

550677 Specifying Control Valve Performance to Achieve Desired Process Performance

In the 1990's, a study of over 5000 control loops revealed that 30% of the loops were INCREASING process variability due to poor valve performance. On a recent control foundation improvement project, over 50% of the loops had poor performance due to significant control valve performance issues. Poor control valve performance continues to be a widespread source of poor control loop performance. A control valve may be properly selected based on physical specifications, but its operational performance may not meet the requirements to achieve the required process performance. Properly specifying valve performance metrics such as dead band, resolution, process gain, step response time, and response regions are vital. Industry standard ISA/ANSI-75.25 provides definitions and test procedures for positioning and response time performance of control valve but does not provide guidance on selecting appropriate control valve performance specifications for a specific application. This presentation explains the impact of valve performance parameters and how to specify them based to achieve the desired process performance. The definition and importance of these performance metrics will be explained, and field examples will be provided. The metrics can also be used to diagnose and improve existing control valve performance when it does not meet the required control loop performance. The benefits include longer valve life, reduced process variability and improved process performance.

549828 Detection and Isolation of Abnormal Event in Nonlinear Industrial Processes By a Novel Data-Based Method

Detection and Isolation of Abnormal Event in Nonlinear Industrial Processes by a Novel Data-Based Method

Chiranjivi Botre[a], Hazem N. Nounou[b], Mohamed N. Nounou[c], and M. Nazmul Karim[a]

[a] *Artie McFerrin Dept. of Chemical Engineering, Texas A&M University, College Station, Texas 77843, USA*

[b] *Electrical and Computer Engineering Program, Texas A&M University at Qatar, Doha, QATAR,*

[c] *Chemical Engineering Department, Texas A&M University at Qatar, Doha, QATAR*

Abstract

Detection and isolation of abnormal events are important aspects in the industry to ensure safe and proper operation of the plant. It is important to accurately, reliably, and quickly detect changes or faults within the process, determine the cause of these changes, and apply corrective actions to mitigate risk. Multivariate statistical techniques are powerful tools that utilize data based model reduction methods capable of efficiently handling process noise and correlated data sets. In literature, multivariate statistical methods are widely discussed for applications in process monitoring and fault detection. Moreover, data based process monitoring techniques have been successfully applied to applications where the accurate process model are not available. Fault detection can be carried out in two phases; data based model reduction and statistical fault detection.

The partial least square (PLS) method is an input-output model that has been effectively applied to linear processes and is most commonly used for fault detection applications. A kernel extension of PLS has been proposed to provide an effective technique for modeling nonlinear industrial processes. In order to enhance the fault detection performance of KPLS model we have developed a new method that optimizes KPLS model through the use of a multi-objective genetic optimization approach. Our multi-scale kernel partial least square (MSKPLS) - based generalized likelihood ratio test (GLRT) method, has the ability to handle process noise, non-normal data distribution, and auto-correlated data sets to provide an effective fault detection technique that can be applied to nonlinear industrial process data.

Once process faults have been detected, accurate isolation of the specific fault within the plant must be achieved to aid swift corrective action. Our approach to fault detection using the advanced statistical GLRT method also allows for multiple faults in the process to be isolated. The isolation method is called a contribution plot and uses the calculated statistical values to determine specific process variables responsible for the process fault, narrowing the possible fault location from plant-wide consideration to the contributions of a key sub-block consisting of a few unit operations. We have developed a novel fault isolation and identification hybrid approach by using a targeted approach where the generalized likelihood ratio (GLRT) based contribution plot is used to isolate the faulty sub-block of a process, and then hybrid model-based observer with a neural network is used to isolate and identify the faulty equipment. We will demonstrate the improved fault detection and isolation algorithm using the nonlinear simulated continuously stirred tank reactor (CSTR) data and Tennessee Eastman Process problem as a case study.

Keywords: KPLS, Wavelet function, Hybrid observers, Fault isolation, Neural network, Continuously stirred tank reactor (CSTR).

References:

1. Bakshi, B. R. Multiscale PCA with application to multivariate statistical process monitoring. AIChE J. 44, 1596–1610 (1998).

2. Botre, Chiranjivi, Majdi Mansouri, Mohamed Nounou, Hazem Nounou, and M. Nazmul Karim. 2016. "Kernel PLS-Based GLRT Method for Fault Detection of Chemical Processes." Journal of Loss Prevention in the Process Industries 43 (September): 212–24.

3. Botre, C., Mansouri, M., Karim, M. N., Nounou, H. & Nounou, M. Multiscale PLS-based GLRT for fault detection of chemical processes. J. Loss Prev. Process Ind. 46, 143–153 (2017).

4. MacGregor, J.F., and T. Kourti. "Statistical Process Control of Multivariate Processes." Control Engineering Practice 3, no. 3 (March 1995): 403–14.

5. Teppola, Pekka, and Pentti Minkkinen. 2000. "Wavelet–PLS Regression Models for Both Exploratory Data Analysis and Process Monitoring." Journal of Chemometrics 14 (5-6): 383–99.

Francis Gregway Shinskey: 2019 and Beyond

Sigifredo Nino, P.Eng.

Summa Control Solutions Inc.

Montreal, QC, CANADA

SNino@SummaControlSolutions.com

Presented at the 2019 Distillation Symposium, AIChE Spring Meeting, New Orleans, Louisiana, March 31 - April 4, 2019

Francis Gregway Shinskey: 2019 and Beyond

Sigifredo Nino, P. Eng. (Snino@SummaControlSolutions.com)

Process Control Consultant, Summa Control Solutions Inc., Montreal, Canada

"Virtually any arrangement of loops will operate satisfactorily in the steady state - but is the unsteady state that tests control performance" (G. Shinskey 1967)

"There is no substitute for process knowledge, and certainly none for common sense" (G. Shinskey, 1994)

Abstract

The purpose of this paper is honour F. G. Shinskey, arguably the greatest contributor to Process Control for the Chemical Processing Industries and, in the process, to inspire the new upcoming generations of process control practitioners to keep his teachings close to their hearts. The paper will present several aspects of Shinskey's life, such as: His most important contributions, His awards, His family farm enterprise, His outspokenness with respect to the academic-industry gap, His musical skills and his paintings, and to narrate where his work and teaching brought him to; amongst other fascinating facts from his life that he was willing to share with this author.

In an age in which filling technical positions is becoming more difficult, engineering seems to have lost its luster (sadly). Interest in engineering from the new generations is waning and the few companies that are still looking for engineers are having a hard time finding them. Process Control expertise is also fading away. It is hoped that this presentation will inform new process control engineers to embrace Shinskey's contributions and instill in them an appreciation to work in the pursuit of a better world. Shinskey was always willing to share his findings and insights. He also encouraged academics and researchers to have a practical and approachable way to their inventions if they wanted to see them become part of the arsenal of tools available to process control engineers in the field. In this day and age, we have a proven engineering discipline; nowadays we need to leverage on the venues offered by the social media and the internet, in general, to carry on the message and timeless techniques Shinskey has gifted to us.

His essential bio

Shinskey was born in Tonawanda, New York on the 29th day of October 1931. He attended the Canisius High School in Buffalo, NY, where he graduated with honors at the age of 16. He graduated Magna Cum Laude as a Chemical Engineer from Notre Dame University (IN) in 1952. He served two tours of duty in Korea while on active service with the U.S. Navy and after working a few years for DuPont and Olin Chemicals, he went on to have a very prolific career at The Foxboro Company from 1960-1993.

While working at Olin Chemicals in Niagara Falls in 1958, he married his secretary Elizabeth Barret, who sadly passed away on December 2018. Eight children were born from his marriage. In Shinskey's own words: "the youngest is a Ph.D. specializing in infant cognition in London

(UK), the other ones live in WV, CA, FL, NJ and two of them in RI, where he as his wife have chosen to spend their last days."

His entrepreneurship

In an e-mail, exchanged on February 16, 2018 for the purposes of this article Shinskey wrote: "I restored a historic (1775) house in Foxborough, and operated the property as a family farm, with a large vegetable garden, raising a steer, pigs, chickens, turkeys, geese, and ducks to feed us year-round. That lasted only as long as we had a houseful to feed."

His artistic inclination

Throughout his lifetime, Greg also had many artistic pursuits. "I was a church choir director and tenor for many years and retired as a church organist at age 80. I began landscape painting with oils, continuing with watercolors and now working with pastels, of which you have some examples." Greg was generous enough to share some of his outstanding art with the author. The Figure 1 shows a couple of his painting, now owned by the author.

Figure 1 - A couple of Shinskey's Paintings

The globetrotter

He wrote a book out of his travel logbook: "TO THE ENDS OF THE EARTH: Travels of an Engineer", which he hoped to get published but it was rejected by the publishers; nevertheless, he wanted to share his experiences as a world itinerant; so herein I quote from the preface: "Though I have written many books, I am not a novelist or journalist, nor well known outside my technical field. By way of introduction, my specialty is Process Control. It is the technology applied to the automatic control of processes in power plants, chemical plants, petroleum refineries, paper mills, etc. I will not bore you with any of the technology; it simply happens to be the vehicle which brought me to the adventures described here. While I am not particularly well-known in my home town, my name is recognized in such faraway places as Australia, Singapore, Venezuela, Finland, and South Africa. In responding to invitations to lecture all over the United States, Canada, and over forty other nations, I lived these adventures."

Having become a lead authority, he was invited to lecture in several famous places such as the Royal Society, the Royal Institute in London, and Oxford University.

The awards

Shinskey's contribution to process control and engineering has been recognized several times, probably not at many as he should have had, probably due to his critical observations of the role of the academic institutions contribution to the preparation of engineers for the real plant challenges. Nonetheless, these are some of the awards he received:

ISA Applications Award (1977) for advances in pH control, ISA Education Award (1983), ISA Founder's Award (1988) for his time domain approach to process control, ISA Fellow (1990), ISA R.N. Pond Award (2003) for best paper of the year, ISA's Life Achievement Award (2008). Computing Practice Award from the AIChE (1992), Benjamin H. Bristol Fellowship, awarded by The Foxboro Company (1982), Sir Harold Hartley Medal, IMC, U.K. (1995), Nordic Process Control Award (1998), Control Engineering Practice Award, AACC (2000), Process Automation Hall of Fame (Control magazine, 2001).

In addition, he holds 17 U.S. Patents for inventions such as "Deadtime simulators for process control apparatus", "Fractionator control systems material balance computer and feedback control", and "Method an apparatus for characterizing and compensating for non-linear components".

The magician-engineer

According to Ed Bristol, Jim Douglas (U. of Massachusetts) spoke highly of Shinskey: "He walks into a room, puts on his black cape and pointed hat, waves his wand at the process and the damn thing works!". The author could not have put it better. His expertise and know-how are nothing short of magic.

Shinskey pioneered the control theory in the time-domain. This occurred after a request to give a course in process control to a heterogeneous group of professionals who would have difficulty following the Laplace domain and frequency domain theories. Those notes eventually became the first edition of his Process-Control Systems book published in 1967. Throughout the years, he expanded and improved this book into three additional editions, with the most recent version published in 1996, by McGraw-Hill in a "Chinese translation of internationally renowned textbooks in information technology and electrical engineering". His books were also translated to Spanish, Italian, Japanese, Romanian, and Russian.

In his time, Shinskey authored well over one hundred papers. His first: "For Gas-Phase Reactors... Design for Control of Temperature", was published in Chemical Engineering, October 5, 1959. And the latest: "Killing Model-Based Control Dead Time", published in Control Magazine in May 2013. If you tally that up, that is over half a century of invaluable contributions to the process control field in the form of articles that provide a wealth of information.

Figure 2 - PCS 4th. Edition in Chinese, 2014

In his time, Shinskey authored well over one hundred papers. His first: "For Gas-Phase Reactors… Design for Control of Temperature", was published in Chemical Engineering, October 5, 1959. And the latest: "Killing Model-Based Control Dead Time", published in Control Magazine in May 2013. If you tally that up, that is over half a century of invaluable contributions to the process control field in the form of articles that provide a wealth of information.

Some of Shinskey's contributions to process control (Not necessarily his inventions)

Shinskey's work in Distillation Control is probably one of his biggest contributions, having written two editions of his DC (1977 and 1982), which happens to perfectly align with this morning's session: Honoring the Contributions of Greg Shinskey: A Jubilee of Advances in Distillation Control.

"Critique of Chemical Process Control Theory" written by Alan Foss in 1973 where he stated that "the central issue to be resolved by the theories of process control is the determination of process control structure", probably, inspired Shinskey to research the subject. In 1983 he presented his Distillation Control methodology to illustrate his proposition on how to address control strategy design problem.

Based on that paper and the collaborations with Shinskey the methodology could be summarized as follows: First, understanding that a model is a simplified and reduced version of the reality, find a process mathematical model; it will open the possibility to better understanding the underlying behavior of the process and will allow for testing that may not even be possible on the actual process. Second, use Relative Gain Array as the principal means for pairing controlled and manipulated variables, supported by the simplification of the size of the multivariable problem provided by the dynamic response of the different process variables, e.g. flows and pressures have much faster dynamics than temperatures and compositions. Although relative gains (RG) can be estimated from plant testing, the availability of suitable differentiable models will allow the RG calculation using slopes instead of steady state gains, such as in

Shinskey's Distillation 4-components proposed method. Third, use partial decouplers, iff (If and only if) necessary. Fourth, apply feedforward to improve controls performance.

Shinskey's contribution spread wide and far among industries such as Heating Ventilation and Air Conditioning, Solar Energy Systems, Mineral Processing, Pulp and Paper, Fossil Power Generation, Oil and Gas, Oil refining, Petrochemical, and food industry. This demonstrates the universality and transversality of control theory and its application to chemical processes that Shinskey was able to exploit masterfully. There are many specific techniques that Shinskey either invented or developed, amongst them, Feedforward control as a high-performance controller (1963), Material Balance in Distillation Control (1963), Separation Factor in Distillation Control (1965), Inferential control (1968), External Reset (1971), Optimizing Control Performance (minimum Integrated Error) by minimizing P, I and Controller output (1972), Values of process control (1974), Development of the Bristol's RG (1967), Override controls (1967).

Shinskey on Energy Conservation – the entropy crusader

Let me begin by clarifying that entropy is NEITHER disorder NOR chaos (Lambert, 2002). Entropy is NOT a driving force, it is an index that measures the energy dispersion within the system and its surroundings; i.e. the distribution of energy among a large number of molecular motions relatable to quantized states. Shinskey has made clear that "energy conservation" is a misnomer, energy is always conserved, but that even if the energy flowing IN and OUT of a process remains equal, its QUALITY deteriorates; to the extent of becoming unusable.

Shinskey wrote (1978, pp.251)" Distillation is a classic example of a process using energy to create order. Typically, the products of a distillation column are in the same physical state as its feed and have the same energy content. Therefore, the energy use in the process simply passes through, increasing its entropy to reduce the of the products". Recalling a conversation where the author asked Shinskey if a highly entropic system would be more difficult to control, he replied "yes". No mathematical proof but an experience-driven statement. Although it was a loose comment, it has a lot of implications when it comes to the possibility of designing and building a process that might be uncontrollable. Process design and control should have the same standing during the design phases of a plant, something Shinskey was also an advocate of.

In the mid-'70s Shinskey became an "entropy crusader". From that period came his proposals, and recommendations, of how controls can be used to save energy and facilitate operation, amongst them, those for Distillation Control: Floating-Pressure Control, Heat-Pump Systems, Compressor-Expander Systems, Regenerative Heat Recovery (Feed Preheat with Bottoms product), Double-Effect Operation (Thermal Integration), and Coupling Unrelated Columns (Energy Integration).

The University-Industry gap

Shinskey was outspoken about the academic-industry gap throughout most of his career. Starting with his 1971 "To teach creativity" writing where he criticized the higher education institutions for their lack of contact with the real world and the consequent continuous proposition of solutions to problems that don't exist in the real life. In 1994, in "A critique of Education in Process Control" he decried the "tendency toward "dumb down" in all areas of education", and on his 2002 "Process Control: As Taught vs as Practiced" he questioned investigation solutions that "has led to research results left unused by industry and graduates left unprepared for industrial assignments", in the same paper he also criticized the study of control loops without nonminimum phase dynamics, overemphasis of setpoint response, the near exclusion of load regulation, and the omission of dynamics in the loop disturbance path.

Is the University-Industry gap closing?

On May 1st, 2018 the following message was posted in the AIChE Engage - Discussion Central:

"Chemical Engineering schools follow a fundamental knowledge approach in educating their students … and I agree that this is the core mission of schools and I see the value in that … The problem is the gap between the fundamental knowledge that fresh grads have and the operational knowledge needed by the industry …"

I have seen strong resistance in academia towards incorporating some kind of operational knowledge/training on systems, equipment, and instrumentation that might accelerate the assimilation of young engineers into industrial operation environments post hiring … I have been teaching chemical engineering topics at the "operational level" for the past two years and I can see the value to the industry in this approach … The question is what does it take to convince "academia" to introduce a "healthy dose" of "operational training" that does not compromise the core mission of "fundamental knowledge-based education" for their engineering students?

Ammar Alkhawaldeh, PhD"

The post received 24 replies in a couple of weeks. So, the academy-operational knowledge divide is still an issue…

His legacy for the New Generations of Process Control Practitioners

Proven Process Control engineering is still needed today. It will also be needed in the future. For newcomers, it is quite relevant as they will be in charge of maintaining control strategies that he/she may have neither designed or have been a part of the implementation or commissioning team. In process control, the logic is backwards: Controls are guilty until proven otherwise; if a unit operation stops working all the sudden, it is the controls fault, even if they have consistently worked fine for years. The process control engineer has to prove that is not the controls by finding out what has really caused the problem. That being said, the engineer will need to be prepared to understand and troubleshoot both, the controls and the process. There is no free lunch on this front!

Education will be required in one way or another, as there will always be the need for in-house expertise in process control. Shinskey's books and papers will always be a source of valuable information whenever there is a need for an ingenious and practical solution to a process control problem. Do not get me wrong, theory is great. Theory allows for the exploration of alternatives not yet seen in the field. But theory cannot be the be-all-end-all to the success of a Process Control Engineer. It must be realistic, and applicable to real life. Keep that in mind.

Based on the commonalities between Shinskey's and my own experiences, as he pointed out on several occasions, the recommendation would be to follow his footsteps. He saw himself as a practitioner who managed to balance theory and practice. He gathered theoretical knowledge from academics like Edgar H. Bristol (1936-) and enriched his hands-on experience from practitioners like Carroll J. Ryskamp (1930-2016), whom he called "Mr. Distillation".

Shinskey always invoked Cyril Northcote Parkinson's law: "The time required to do a job will be equal the time allowed to do it" So make sure that you set reasonable limits to the delivery of your work, otherwise the work will never be finished, and the managers will not see the value of your contribution as process control engineer.

On Artificial Intelligence and other tools

The author wrote an expert system based on backward reasoning back in 1987. An expert in the field, a knowledge engineer and a computer programmer were required to build an expert system. To avoid all the risks associated to dividing the work among people from different disciplines, Shinskey himself wrote a Distillation Expert System back in the mid 80's. The author used Shinskey's tool: it provided help into getting to know what it takes to design distillation control strategies. Yet, there were cases where the software application could not be used as the column specifications would not fit the models and/or assumptions made in the expert system, it other words, even if it had the essence of Shinskey's expert knowledge of the subject it was not a substitution of him, by any stretch of the imagination. As a matter of fact, it was only when the fundamentals of distillation column design, operation and the understanding or underlying principles behind the expert system were grasped, that the use of the expert system became more "normal" to this author. Bottom line, a tool is not a substitution for either experience or knowledge.

In 2018 this author took an eleven weeks course on "Machine Learning and Data Mining"; software tools have remarkably improved over the past 30 years, however, the same "limitations" apply: "No magic (except for Shinskey's!), No free lunch". The data has to be either collected or extracted from a historian for what the purpose of the application is, the data has to be adequately selected, which can take 70% to 80% of the project time, the data mining algorithms must be understood, the resulting models need to be monitored and recalibrated, etc. All these is to give a clear message to the new generations of process control engineers: Tools are tools, and are not substitution for human expertise. Theory-Praxis Feedback in the learning process is priceless.

On persistence

The author's work with Shinskey for over 25 years, gave him the opportunity to learn from the man himself. At one facility, after years of the continued operation of a successful application of his material-balance controls on two heat-integrated distillation columns (Nino, 2011), the author was told by the primary contact at the facility, that nobody at that plant thought that it would work! No sure if he ever received that type of comment from any customer.

Shinskey firmly believes, as the author does, and as any process control engineer must know: There are lots of value and benefits to be accrued by applying sound process control strategies to the chemical processing plants. As far as Distillation Control is concerned: Shinskey told the author during the execution of a sidestream drawoff distillation column project: "Give material-balance controls a chance".

The moral of the story is: If you have a solid design, backed by strong engineering fundamentals, the trends on the DCS will speak on your behalf. All you need is the opportunity from the right person to prove the benefits or process control. Sadly enough, this is more often than not the hardest piece of the puzzle to solve. Finding a believer, in a position of authority is difficult. For this reason, persistence is a must-have in the Process Control Engineer toolbox.

References

Felton B. (2008) "Innovating, educating, writing, earn Life Achievement Award for Greg Shinskey", Intech Magazine, August.

Lambert F. L. (2002) "Disorder – A Cracked Crutch for Supporting Entropy Discussions", Journal of Chemical Education, Vol. 79 No. 2, February

McMillan, G. and S. Nino (2017) "The greatest source of process control knowledge", Control Magazine, October

Shinskey, F. G. (1967) *Process-Control Systems*, McGraw-Hill Inc., NY.

Shinskey, F. G. (1978) *Energy Conservation Through Control Systems*, Academic Press, New York (NY)

Shinskey F. G. (1979) "Entropic and Anti-Entropic Processes and the Future of Technology", Energy, Man and Civilization - Report on an Engineering Foundation Conference. Engineering Foundation, NY.

Shinskey, F. G. (1971) "To teach creativity", Instrumentation Technology, pp.34-38, October

Shinskey F. G. (1983) "From process knowledge to computer control: A methodology", Pacific Area Chem. Eng. Congress, Seoul, Korea, May 5-8

Shinskey F. G. (1994) "Process Control: Where have we been, where are we going", South Africa IMC symposium, Durban, June.

Shinskey F. G. (1994) "A Critique of Education in Process Control", IASTED Conference on Computers in Industry, Cairo, December

Shinskey F. G. (2002) "Process Control: As Taught vs as Practiced," Industrial & Engineering Chemistry Research, 2002, vol. 41, no. 16, pp. 3745-3750

DEEP UNDERSTANDING AND GREAT ACHEIVEMENTS OF PID CONTROL

Gregory K. McMillan, Emerson Automation Solutions, Chesterfield, Missouri

Biography

Greg retired from Solutia Inc. as a Senior Fellow after 33 years specializing in improving loop performance. Greg contracted as a consultant in DeltaV R&D for 12 years and is presently a Principal Senior Software Engineer in Simulation R&D for Emerson Automation Solutions focusing on dynamic modeling using the Digital Twin. Greg is the author of 80 articles and papers and 20 books. Greg's most recent books are *Advances in Reactor Measurement and Control*, *Tuning and Control Loop Performance 4th Edition*, *Good Tuning: A Pocket Guide 4th Edition*, and *Process/Industrial Instruments and Controls Handbook 6th Edition*. Greg is an ISA Fellow and received the ISA "Kermit Fischer Environmental" Award for pH control in 1991, the Control Magazine "Engineer of the Year" Award for the Process Industry in 1994 and was inducted with Greg Shinskey into the "Process Automation Hall of Fame" in 2001. Greg was honored as one of InTech's 50 most influential innovators for advancing automation and control technologies and received the ISA Life Time Achievement Award in 2010.

Abstract

Shinskey's articles and books are the best source of what is truly important and achievable by PID control. His extensive analysis of process relationships and dynamics and PID controller capability are unmatched and is the key to process control systems that can increase process capacity and efficiency. He has been the biggest influence in my career. Many of accomplishments can be traced back to knowledge gained from Shinskey.

Process metrics such as integrated error and peak error can readily be estimated for disturbances from the controller tuning settings showing the impact of poor tuning. The disturbances are not on the process output as what is commonly shown in publications but are on the process input. The implications in terms of focusing on what is truly important and eliminating distractions from special algorithms supposedly better than PID depend upon this fundamental understanding. PID feedback and feedforward tuning settings and tests should be based on recognition of the location and dynamics of the disturbance. Any desired setpoint response can be achieved by the simple addition of a setpoint lead-lag. The use of external-reset feedback can automatically prevent problems with cascade control, poor control valve response and the use of analyzers with variable sample or cycle times. Furthermore, dead time compensation can be more effectively achieved and its limits recognized. Simulation can be used to develop the best automation system and control strategies. The benefits gained from the insights should continually grow. It is our job to make sure this is the case.

Introduction

There is no way to convey the incredible body of knowledge developed and documented by Shinskey. Nearly all of the knowledge stems from a deep understanding of the process, dynamics, PID capability, and the role of tuning. His publications address and eliminate the many misconceptions that unfortunately still prevail. If there was just one wish to be granted by some magical power, it would be that every academic and practitioner would know and use what Shinskey knew.

Sample of Many Great Achievements by Shinskey

The sample of key learning points offered are indicative of fundamental principles that are essential to improving the performance of control loops for nearly all types of processes. Understanding these points can get us all going in the right direction. This can lead to much greater recognition of the value of control and optimization in improving process performance.

Power of external-reset feedback to reduce oscillations

By simply turning on the PID option for external-reset feedback (e.g., dynamic reset limit) and sending back to the PID a fast update of the actual process variable, valve position or speed being manipulated, the PID output from integral action will not change faster than the secondary loop, valve or variable frequency drive (VFD) can respond.

This capability reduces oscillations from valve backlash and stiction, poor positioner sensitivity, slow actuators, VFD setup deadband and rate limiting, violation of cascade rule (primary loop 5 times faster than secondary loop) and interaction. It can improve batch control and override control. It also provides directional move suppression by the simple addition of PID or analog output setpoint up and down rate limits to prevent unnecessary crossings of the split range point, fast opening and slow closing surge valves, less disruption from high PID gain action, and smooth gradual optimization by valve position control with a fast getaway for abnormal conditions.

Dead time compensation that is much simpler

By simply inserting a dead time block in the external-reset feedback path back to the PID, you can achieve dead time compensation as effective as a Smith Predictor but without the need to insert an open loop process gain or time constant into a model. Also, there is no loss of the process variable in the operator interface and the dead time can be updated as function of process operating conditions such as production rate. Shinskey in his tests eliminated many widespread misconceptions such as the benefit being greatest for dead time dominant processes (greatest benefit is for lag dominant processes), the PID must be more aggressively tuned to see the benefit, the compensator can do better than MPC, and the dead time compensator is more sensitive to a block dead time larger than actual dead time. In a regular PID controller, larger than actual dead time used in tuning does not create oscillations.

Integrated error metric as a function of tuning, scan time and filter time

Missing in nearly all of the literature is the key fact that the integrated error for a load disturbance is a function of tuning settings. Studies showing the benefits of algorithms developed to replace the PID typically fail to take this into account allowing the developer of the algorithm to tailor test results to show the supposed benefits of the new algorithm. The equation developed by Shinskey is incredibly elegant. Shinskey extended it to include the effect of scan time and filter time by simply adding them to the reset time in the numerator.

Realization that most disturbances affect a process input

Perhaps the greatest disconnect between academia and industry stems from not realizing that most disturbances are on the process input instead of the process output. The consequences are greatly increased by studies that use balanced self-regulating processes where the primary process time constant is about the same size as the process dead time. The result has been algorithms and tuning methods that would be detrimental to the point of being hazardous for many of the more important composition and temperature control loops in the chemical industry. These loops tend to have a near-integrator or true integrating process response and in some cases, a runaway response demanding a much more aggressive algorithm and tuning. For example, some experts whose world was only balanced self-regulating loops advocated tuning to prevent overshoot by the controller output of its final resting value. This would be unsafe for near-integrating, true integrating and especially runaway processes. Shinskey has written about open loop unstable processes (e.g., parallel heat exchangers, steam jets, and highly exothermic batch polymerization reactors).

Tuning first for good load response and then good setpoint response

Nearly all tuning methods tested by making a setpoint change when in fact the tuning objective should in most cases be tuning for dealing with unmeasured load disturbances. The test for such disturbance is quite straightforward and simply involves momentarily putting the controller in manual, making an output change and putting the controller back in auto. The tuning should minimize integrated and peak error with minimal oscillation. Good setpoint response can be obtained by inserting a lead-lag on the setpoint. The lag is simply set equal to the reset time and lead is set equal about ¼ the lag time. Using a lead helps get the process variable to setpoint faster. If this is not a requirement (e.g., bioreactor pH shifts), the lead time can be left at zero, which is equivalent to a PID structure of PD on PV and I on E.

Controlled and manipulated variable pairing

The best pairing is normally the controlled and manipulated variables with a relative gain array value near unity. For integrating processes, the process variables are converted to a variable with a steady state being a constant ramp rate by simply using the rate of change of the process variable. Digital Twin simulations are good at finding the relative gain array. Testing in the field should be done to confirm choices. Sigifredo Nino, Shinskey's protégé is excellent in using RGA particularly for two point distillation column control. Shinskey developed many other criteria based on first principles, such as reactor endpoint control. A reactor pressure controller manipulating a gaseous reactant flow for a liquid co-reactant and liquid product and a reactor level controller manipulating a liquid reactant flow for a gas co-reactant and gas product will inherently manipulate the reactant in the right stoichiometry.

Dead time from discrete updates

There are a multitude of studies on the effect of I/O scan time, PID execution rate, wireless update rate and analyzer cycle time. The results are usually not truly representative and cannot be extended to other cases. The analysis could be simplified to the inherent effect simply being the ½ the time between updates plus latency. For scan times and wireless update rates, the latency is zero. Thus, the dead time is simply 1/2 the scan time or update rate. For PID execution rates, the latency can be significant if there is a lot of computation going on between the analog input and output besides the PID. For analyzers where the result is computed at the end of the cycle, the latency is the cycle time. The resulting dead time is then 1½ times the cycle time. Of course, there is also the sample transportation delay.

Simulation can quantify PID and dead time compensator performance

Shinskey extensively used simulation to show PID performance and robustness was affected by tuning and dead time compensation for various types of dynamics. This experimentation helped dispel many myths and verify the value of the PID and inherent tradeoff between performance and robustness. this concept is tended by modeling on a first principle basis challenging processes including automation system dynamics (e.g., delays, lags, deadband, resolution, sensitivity ...). For pH, Shinskey's charge balance is extended to include conjugate salts to detail the equipment and piping besides PID design requirements. Most recently, the Digital Twin is used to eliminate the errors associated with emulation and make the exchange of control system configuration and operator graphics seamless.

Block Diagram Showing External-Reset Feedback

Pneumatic controllers had to implement integral action by the use of a bellows and restrictor creating a filter whose time constant was the reset time in what is a positive feedback implementation of integral action. When electronic and digital controllers were developed most suppliers decided integral action could now be done with an integrator as seen in textbooks. Shinskey knew better and made sure Foxboro kept the use of a filter with a positive feedback implementation. Since most systems used a Series PID Form, his books showed this implementation. The time domain block diagram shows this capability along with eight different structures in a Parallel PID Form in DeltaV given as the default "Standard Form". A key to understanding this Form is that that when the error is zero, the input and output of the filter are equal suspending positive feedback. Also, integral action cannot be faster than the actual response of manipulated loop, valve or VFD in the external-reset feedback of the manipulated variable (MV) whether a process variable, stroke or speed. If the MV is not responding, integral action is suspended. Showing how to implement the parallel PID for a setpoint weight factor (β) requires numerically backing it out in the contribution to the output.

Examples Where External-Reset Feedback Helps

I find more and more situations where external-reset feedback suppresses oscillations or enables a more intelligent PID action. The fundamental capability by simply turning on this option is largely overlooked as seen in the fact that the PID algorithms in most DCSs and PLCs do not have the positive feedback implementation of integral action that gives external-reset feedback as an inherent capability.

Slow secondary loop in cascade control

While it is certainly better to design a cascade control loop so that the secondary loop is at least five times faster than the primary loop, turning on external reset feedback can reduce oscillations as seen on slide 5 for a setpoint change to the primary loop. Undershoot is reduced from 5.2% to 1.5% and the settling time is reduced from 1200 to 275 seconds.

Limit Cycle from backlash (deadband) in system with two or more integrators

A limit cycle will develop from backlash (deadband) if there are two or more integrators in the automation system and the process whether due to cascade control, a positioner with integral action, or an integrating process (e.g., gas pressure, batch composition or temperature, or level). Deadband can also exist from configuration in split range blocks and in VFD setup to reduce the response to noise. Turning on external reset feedback can slow down and the reduce amplitude of the oscillations in the process variable as seen in slide 6 for level.

Limit Cycle from stiction (resolution) in system with two or more integrators

A limit cycle will develop from stiction (resolution) if there are one or more integrators in the automation system and the process. There is almost always at least one integrator and the use of valves and digital systems ensure there is a resolution limit. Consequently, there is always a limit cycle albeit quite small for good throttling valves and a modern DCS or PLC with properly spanned transmitters. The standard I/O cards for a VFD speed setpoint and older DCS thermocouple and RTD I/O cards can have poor resolution creating a noticeable limit cycle. Turning on external-reset feedback can stop the limit cycle for a single point of integration but with an offset as seen in slide 7 for a flow loop.

Slow valve slewing rate or VFD speed rate of change limiting

Large actuators without a volume booster on positioner output and VFDs with a speed rate of change limit introduced due to an over concern of motor overload create a problem not seen for small changes in PID output. There is a burst of oscillations for large changes. Turning on external-reset feedback suppresses the oscillations as seen in slide 8 for a valve.

Poor positioner sensitivity (response time increases by order of magnitude)

Most people are unaware their positioner has poor sensitivity because tests normally use changes in signal of 1% or more. If the signal change is less than 0.5%, many positioners develop a response time that increases by a factor of 10 or more (e.g., 40 instead of 2 seconds). The response time also tends to double for a reversal of direction (e.g., 80 seconds). The solution is to turn on external-reset feedback of a fast readback of actual valve position and increase the PID gain to make larger changes in the PID output seen in readback.

Fast opening and slow closing surge valves

The addition of up and down setpoint rate limits to the analog output block can enable a fast opening of a surge valve to prevent surge and a slow closing to ensure a more gradual approach to more optimum operating conditions. In the past this was done in the field, with limited adjustability and visibility. The controller has to be tuned for the slowest response. By using external-reset feedback, tuning for fastest response can be used and the setpoint rate limits can be visible on the operator interface and accurately set in the PID configuration.

Unnecessary split range crossings

The discontinuity and nonlinearity at the split range point tends to make loops oscillate at this point resulting in detrimental consequences in terms of variability, loss of efficiency and in the case of bioreactor pH control, high osmotic pressure causing cell death from an accumulation of sodium ions form the unnecessary addition of sodium bicarbonate. Patience can be achieved by adding a setpoint rate limit slowing down a change in the use of a utility or reagent in the direction of with the least safety and process performance risk. External-reset feedback of the actual manipulated variable prevents the PID form continuing to drive the setpoint in the less desirable direction. This extra patience reduces reaction to short term situations and disruptions at the split range point.

Slow approach to optimum and fast getaway in valve position control

The addition of setpoint up and down rate limits provides adjustable and visible directional move suppression. For valve position control that maximizes feed rate or minimizes utility use by pushing user valves to a maximum or minimum throttle position, move suppression in the direction of increasing feed rate or decreasing utility use is done smoothly and gradually. However, if there is load upset or some undesirable situation, the move suppression in the opposite direction is made much faster to provide a fast getaway. External-reset feedback enables directional move suppression without the need to retune the PID.

Upset to other loops due to aggressive gain setting

In many applications the more immediate response of proportional action is needed to deal with load disturbances for process safety and performance. This is especially the case for near-integrating, integrating and runaway processes (e.g., composition, gas pressure, and temperature control). However, rapid movements of the manipulated variable are upsetting to the operator and can be upsetting to other loops particularly those sharing the same header or resource. The addition of setpoint rate limits to the analog output block or secondary loop with external-reset feedback of the actual manipulated stroke, speed or process variable can be optimized to meet loop objectives without upsetting other loops or the need to retune the PID.

Preload for best batch operation setpoint response

The time to reach setpoint without overshoot can be optimized by a smart preload response that puts the PID output at intelligent positions during the setpoint response providing a smooth response from P or PD to PID control.

Smooth and intelligent transition to selection in override control

The filter in external-reset feedback smooths out disruptions from noise, discontinuities and nonlinearities in override control. The unselected PID output integral mode contribution gradually moves to match the manipulated variable. This provides an ability to understand and optimize the contribution of proportional mode to provide the best response for selection.

Load Performance Visual Actualities

Shinskey showed that the optimum response to a load disturbance is simply a triangle where gains smaller than ultimate limit cause a higher triangle peak and slower reset times causes the ride side of the triangle to develop a slow tail toward zero error. This realization is particularly useful to switching to integrating tuning rules for near integrating processes and setting an arrest time that can be as small as 1/2 dead time for fastest load response.

Load Performance - Practical Limits from Tuning

Shinskey provided an insightful equation for integrated error for an unmeasured step disturbance. The numerator is simply the sum of the PID reset time, scan time, and filter time. The denominator is the controller gain. The change in PID output required is converted to an open loop error that puts the open loop gain in the denominator. An equation for peak error is possible starting with the one by Peter Harriott for a proportional only control and then adding an approximation of the slight effect of integral action.

Load Performance - Ultimate Limits from Dynamics

Often we lose sight of the ultimate limits imposed by process dynamics, particularly the total loop dead time. The peak error is the ratio of the loop dead time to the summation of the loop dead time and largest lag in the loop ideally set by a large back mixed process volume. The integrated error has the dead time squared in the numerator. These equations emphasize the importance of minimizing dead time.

Sources of Dynamics

Academic publications present a linear idealized view of process dynamics. The result is a loss of understanding of effect of automation system dynamics that for many loops limit the performance. These dynamics are the responsibility of the process control engineer and the ability to make improvements are faster and less expensive than changes to equipment and piping. The total loop dead time is the summation of all the pure delays and the fraction of time constants in series that become equivalent dead time in a first order plus dead time approximation. The largest time constant is hopefully in the process to slow down the process excursion from a load disturbance. However, for many liquid pressure and flow control loops, the largest time constant is in the automation system from valve or measurement response or process variable filter. A secondary time constant can be singled out to provide a second order approximation. The conversion of a near-integrating process gain to an integrating process gain is noted as simply the process gain divided by the process time constant. The open loop gain is the product of the valve or VFD gain, process gain and measurement gain showing the effect of valve and pump sizing and measurement scale. The resulting open loop gain is dimensionless for self-regulating processes and 1/sec for integrating processes. Many people do not realize that the PID works in percent of signals and not engineering units.

Dead time from Measurement System

There are many sources of the measurement dead time including transmitter update rate, analyzer cycle time, sensor lag(s), sensor transportation delay, transmitter damping, and resolution or sensitivity limit that is the limit divided by the process variable rate of change. The dead time from valve backlash and stiction can similarly be estimated as the dead band or resolution limit divided by the rate of change of controller output.

Use of Digital Twin to Become More Like Shinskey

We have incredible untapped capability in today's measurements, control valves, controllers, and software. What we are missing is the recognition of what Shinskey's knowledge and our systems can do to increase the performance of the operator and plant. The Digital Twin can reveal the opportunities, detail the solutions, and identify and achieve the benefits that can lead to a rejuvenation of our profession providing the motivation utilizing and building on what we learned from Shinskey. Detailed first principle dynamic models with automation system dynamics added can reveal the actual performance revealing the consequences of process and mechanical design, control strategies, procedure automation, and PID options and tuning. The powerful capability of new blocks and modeling objects enables the average process control engineer to fully exploit the opportunities and give the knowledge gained to all of the personnel running and supporting the plant. The Digital Twin has offers technical fluency and skill in addressing process control improvement opportunities in pH, chemical reactor, bioreactor, and compressor control. The key roles are emphasized in exploring=> discovering => prototyping => testing => justifying => deploying => testing => training => commissioning => maintaining => troubleshooting => auditing => continuous improving showing "before" and "after" benefits of solutions from online metrics. It is important to note that while we have singled out process control engineers in this presentation, the need for knowledge to attain the best performance extends to operators, maintenance technicians, process engineers, mechanical engineers, and information technology specialists (e.g., data scientists). Just think what can be realized if we were all on the same page understanding the process and operational opportunities and the value of the best measurements, valves, strategies, and PID?

Digital Twin's Key New Blocks

Blocks have added to the Digital Twin to test innovations and improvements in strategies, dynamics and PID tuning. The PID Performance block computes the integrated and peak error for load disturbances and the rise time, settling time, overshoot and undershoot for the setpoint response. The block is automatically triggered by a change in PID setpoint or output. A Backlash-Stiction block can be put in series with each other to simulate what is happening in the actuator, shaft to stem connection, and stem to ball or disk connection. Blocks to include a variable dead time, slewing rate and response lags can be inserted. The Future Value block can provide a timely low noise rate of change and future value typically a factor of the total loop dead time into the future from current value. This is extremely useful for batch endpoint prediction, profile control, procedure automation, surge curve and actual surge identification, RCRA pH violation prevention, and impending Safety Instrumented System activation.

Conclusion

This is just a brief insight into the incredible deep understanding gained and great advances achieved by Shinskey. Reference 2 below includes a list of nearly all of the articles and books by Shinskey. We should read and learn and use the wealth of knowledge given by Shinskey.

References

1. McMillan, G.K. and Vegas, P. H., *Process/Industrial Instruments and Controls Handbook Sixth Edition*, McGraw-Hill, 2019 https://www.mhprofessional.com/9781260117974-usa-process-industrial-instruments-and-controls-handbook-sixth-edition [mhprofessional.com]
2. McMillan, G.K. and Nino, S., "The greatest source of process control knowledge", *Control*, Oct 2017 https://www.controlglobal.com/articles/2017/the-greatest-source-of-process-control-knowledge/?stage=Live
3. McMillan, G.K. and Berutti M.E., "Virtual plant virtuosity", *Control*, April 2017 https://www.controlglobal.com/articles/2017/virtual-plant-virtuosity/?stage=Live
4. McMillan, G.K., "How to specify valves and positioners that don't compromise control", *Control*, May 2016 https://www.controlglobal.com/articles/2016/how-to-specify-valves-and-positioners-that-dont-compromise-control/?stage=Live
5. McMillan, G.K., *Tuning and Control Loop Performance, Fourth Edition*, Momentum Press, 2014 http://www.momentumpress.net/books/tuning-and-control-loop-performance-fourth-edition

551691 Importance of Regulatory Control Design for MPC

Greg Shinskey has had a tremendous impact on the practice of process control engineering going back to the late 1970s and 1980s. During this time frame, the arrival of the Distributed Control System (DCS) and the minicomputer allowed Shinskey's process-oriented, advanced regulatory controls to be readily implemented. Taking further advantage of increasing computing power, Model Predictive Control applications came on the industrial scene in the mid 1980s, with significant growth coming in the 1990's and thereafter. MPC provided built-in capability to handle process/control interactions, constraints, and optimization, ostensibly removing the need to improve or modify the regulatory design.

With the benefit of experience gained over many MPC applications, the recognition of the importance of the regulatory control design for MPC has only increased, and an acknowledged best practice is to jointly design both the MPC and regulatory controls. Design considerations include disturbance rejection, constraints, and optimization objectives. As a result, many of the techniques and reasoning behind Shinskey's approaches to process control hold true today.

In this presentation, we will highlight, via examples primarily involving distillation columns, the considerations that go into the design of the regulatory controls and MPC, while making reference to Shinskey's contributions.

551693 The Forgotten Science of Distillation Control

Distillation is the heart of all refining and petrochemical plants, as it determines the ability of a plant to

generate on specification product(s). In spite of the vast body of fundamental research and practical

design experience within the industry, the design and tuning of the regulatory controls, which plays a

pivotal role in achieving on specification production, became and remains largely a lost and dying art.

This is evidenced by the author's experience across multiple sites and companies wherein similar, if not

identical, controls design and control ability issues were discovered and rectified on and around

distillation units, as well as by the precipitous drop in attendance of control related sessions at industrial

and academic conferences.

Historically, there was a great amount of attention given to the design and performance of the

regulatory and advanced regulatory controls (ARC) implemented on distillation units. As more advanced

technologies came into common use, less attention was given to the underlying regulatory control

structures, in spite of the fact that it provides the foundation upon which the new technologies rely. IN

parallel to the shift in technology usage, the past decades have seen fundamental changes in the modus

operandi of refineries: the design and operations resemble that of chemical plants. This was primarily

the result of end user quality programs that propagated more stringent product specifications

"upstream" in the supply chain. Despite the changes to the operational criteria these tighter

specifications created, refinery designers have effectively maintained the historical "status quo",

especially regarding controls design.

With these technological and historical aspects in mind, this paper presents fundamental and practical

guidelines for designing robust and reliable control structures on these vital units. The authors draw from 30 years of experience in process and controls design, and tuning such controls, on multiple units

across a wide array of refining, petrochemical and LPG/LNG plants. In addition to the guidelines, a

sampling of real world situations and the solutions for same will be presented to illustrate how the

principles are practically implemented. Topics included in the presentation will include:

- Only Two "Handles" for Every Distillation Tower
- Accumulator Level Control Myths
- Relative Gain Analysis Made Easy
- "Closing the Loop" in Controls Design
- When Tray Temperatures act like Levels
- Flooded Condenser Design: Keep it Simple!
- APC: Love it, Hate it, Need it?

549967 Controlling Trace Impurities in a Dividing Wall Distillation Column

Though attractive due to capital and operational savings, dividing wall distillation columns (DWCs) have yet to gain widespread industrial acceptance. One notable concern is controllability. The research in this presentation investigated the operational flexibility of a fixed-design DWC using a four component mixture, tested the ability of conventional distillation control design techniques to determine control structures for a DWC, and created a minimum energy operating surface that could be used to examine control structures.

As a form of process intensification, DWCs have a higher risk for controller interaction rendering conventional PID control potentially inadequate. This work successfully used two PID temperature controllers to maintain the column at steady state, transition the column between steady states, and reject feed disturbances without controller interaction. These controller pairings were determined using conventional controller design techniques. Therefore, for this chemical system and column design, traditional approaches to distillation control are sufficient to handle the intensified nature of DWCs.

Because more components are present in DWCs in larger amounts, there is concern that temperature control will no longer imply composition control. Temperature control proved successful in this study. Controlling two temperatures maintained column operation against feed disturbances. In addition, prefractionator temperature correlated well with reboiler duty for multiple feed qualities therefore serving as a promising control variable. A steady state model with heat transfer matching the pilot data was scaled to the size of an industrial tower and used to generate a minimum energy response surface for different vapor and liquid split values.

Introduction

DuPont™ IsoTherming® Hydroprocessing technology is emerging as a cost-effective alternative to the traditional trickle-bed reactors in refinery hydroprocessing applications. Performances of many hydrotreating and hydrocracking catalysts have been observed in a pilot plant setting as well as in commercial operation. Most of the catalysts are bimetallic and contain nickel or cobalt with molybdenum or tungsten on an alumina support for hydrotreating applications and on a zeolite support for hydrocracking applications.

In the refining industry, highly aromatic light cycle oil (LCO) offers a limited market value due to its low cetane value, high density, and high sulfur content. LCO must undergo hydrotreatment in order to meet diesel product specifications, but the highly reactive nature of the hydrocarbon stream causes complications with regards to supplying the necessary hydrogen and managing the considerable exothermic heat release. There are commercial units using conventional trickle-bed technology to process the LCO with the use of hydrotreating and hydrocracking catalysts to meet product specifications, but often at a significant yield of the primary diesel product.

Materials and Methods

Continuous pilot plant tests were conducted using a bench-scale reactor system containing 1" O.D. by 20" long flow reactors. Temperature, pressure, and gas & liquid flows in this continuous pilot unit were controlled using a PLC system. Multiple reactors packed with commercial bimetallic hydroprocessing catalysts, each reactor corresponding to a catalyst bed in a commercial reactor, were used in the pilot tests. Business goal was to conduct several weeks long tests using a customer supplied LCO feed to obtain the kinetic information for preparing a commercial licensing proposal. In a typical test, the commercial catalyst was dried, sulfided and stabilized (pre-coked) in the first week. One to two data points were then obtained per day in the remainder of the test at different temperature, pressure, and liquid/gas flow rates. Feed and product samples were analyzed to obtain sulfur & nitrogen removal kinetics, as well as the chemical hydrogen consumption rate and the aromatic saturation level. Feed and total product samples (TLP) were also distilled to obtain the relative naphtha and diesel yields.

Results and Discussion

IsoTherming® technology offers a one- and two-stage process to hydrotreat highly aromatic LCO using commercial bimetallic hydrotreating (HT) and hydrocracking (HC) catalysts to meet the desired product specifications while also maximizing product yield. While commercial trickle-bed processes often result in diesel yields of about 50-60%[1], a one-stage IsoTherming® process, consisting of both HT and HC catalysts under a single liquid recycle, have resulted in diesel yields of up to 80%[2]. The efficacy of the zeolitic cracking catalyst may be further increased and yields greater than 90% achieved in a two-stage IsoTherming® process [3] [4] [5] [6], which includes a separation step between the hydrotreating and hydrocracking sections to remove dissolved ammonia that has a poisoning effect. The intermediate separating step could be a simple high pressure flash alone, or in combination with a stripper or a fractionator. The value of the desired product plus the relative magnitude of the associated capital expenditures, the reduction of the product yield loss and the utility savings, determine the most economical option.

This presentation will discuss pilot plant tests of LCO hydroprocessing in IsoTherming® reactors using commercial, bimetallic hydrotreating and hydrocracking catalysts including tests with and without a separation step after hydrotreating (as is the case of a two-stage and one-stage process, respectively). The examples will demonstrate the principles and the advantages of the liquid-full IsoTherming® reactors over the traditional trickle-bed reactors used in refineries today.

Significance

Liquid-full IsoTherming® reactors offer advantages over the traditional trickle-bed reactors for hydroprocessing of oil refinery streams. The advantages and the basic operating principles of the IsoTherming® reactors are demonstrated for Light-Cycle Oil (LCO) upgrading for maximum diesel production.

References

1. Thakkar, V.P. et al. LCO Upgrading, AM, 05-53, NPRA, p. 1-15, 2005.
2. Dindi, H., et al. Hydroprocessing light cycle oil in liquid-full reactors, US 8,721,871 (2014)
3. Dindi, H., et al. Targeted Pretreatment and Selective Ring Opening in Liquid-full Reactors, US 9,139,782 (2015)
4. Dindi, H., et al. Hydroprocessing light cycle oil in liquid-full reactors, US 9,139,783 (2015)
5. Dindi, H. et al. Liquid-full Hydrotreating and Selective Ring Opening Processes, US 9,212,323 (2015)
6. Dindi, H., et al. Hydroprocessing light cycle oil in liquid-full reactors, US 9,139,783 (2016)

550652 Optimising Chemical Cleaning of Heat Exchangers

Chemical cleaning methods provide a flexible, economic and quick way of restoring heat exchange performance. However, a wrong selection of the cleaning agent and/or duration of the cleaning action can lead to uneconomical decisions. In this paper, a framework integrating condition-based cleaning models that capture kinetics and efficiency of a cleaning action as a function of the characteristics of fouling deposits (e.g. composition, layering, degree of coking) with a predictive maintenance simulator is presented. The approach uses a few parameters, determined in the lab, to predict cleaning performance of given solvents thus providing the field engineer with actionable information on the best method to be used, duration of the cleaning and expected economic benefits. Results are presented using case studies for individual heat exchangers and entire networks.

551768 The Contingencies of Distillation Columns for Ers Design and Special Reference to the Location of the Relief Device

This presentation emphasizes the unique contingencies applicable in the emergency relief system design of tray and packed columns used for mass-transfer operations. Attention is given to the internals of the column and their impact on such phenomena as flooding and obstruction to the flow required to attenuate pressure during an emergency relief. While the relief devices should always be located at top of the column to promote unidirectional vapor relief flow and process vapor flow thereby avoiding reverse vapor flow of relief and consequential damage of internals, there are circumstances when the flow of the relief device(s) located at the top of the column is high enough to cause flooding. An unwise mode of differential pressure control with coincident double jeopardy of overhead condenser failure may also create high relief load. The flow may be obstructed due to softening of non-metallic packing material under a fire scenario, coking, or formation of pop-corn polymers. Too high a flow may cause a pressure drop through a tray that may damage the tray. Under these circumstances the choices are consideration of metal packing, redesign of tray, tray support, and an inherent means of reducing flow through relief device by restricting lift or restricting the scenario flow, safety instrumentation, or location of a relief device at the bottom section of the column in the vapor space. If the location of the relief device at the bottom of the column is unavoidable, two-phase flow relief along with phase separation in effluent handling equipment and safe relief discharge location must be considered. This is a price that a company must pay for irresponsible project and process management in not engaging an Emergency Relief System (ERS) design Engineer in the conceptual and design phases of the engineering.

In the end, an ERS design of a mass-transfer column without the consideration of the internal is not only hollow, but also equally irresponsible.

551769 A Practical and Cost-Effective Solution to Monitor Product Recovery Rate in High Purity Side Draw Distillation Columns: An Industrial Methanol Distillation Case Study

The measurement of product recovery ratio in industrial methanol distillation is of high economic importance and represents a key performance index (KPI) for the distillation unit. However, in many plants the product recovery of the distillation units is not actively monitored. Instead, it is back calculated based on daily production figures. This is primarily because the active monitoring of product recovery can be costly as this requires the installation of expensive gas chromatographs and accurate feed mass flow measurements, which can be difficult to justify economically. In this work we have developed a simple and economical method based on density and flow rate measurements to calculate the product recovery of industrial methanol distillation columns. This method has been validated against plant measurements as well as a validated process simulation. Step and disturbance tests carried out suggest the proposed method is able to accurately estimate the product recovery within the plant operational envelope.

544144 Multibatch Distillation Systems (MultiBatchDS™) and Novel Batch Configurations

Batch distillation is an unsteady-state operation. It is because of this unsteady-state nature that batch distillation columns can be operated in multiple configurations, multiple operating modes, and can have multiple fractions or cuts. The new column configurations like the batch stripper and the middle vessel columns can circumvent the steady-state mass balance restrictions of continuous columns resulting in better purity and better yields.

Multibatch distillation systems (**MultiBatchDS™**) is a software package which captures the essence of these multiple configurations, multiple operating modes, and various fractions of batch distillation by providing the capability to simulate, design, optimize, and control such columns with multiple levels of models.

This talk presents few real-world case studies different operating companies, demonstrating the usefulness of new column configurations of batch distillation.

551770 Details of Industrial Batch Distillation

Batch distillation is one of the oldest separation processes. It is still widely used in chemical industry due to its flexibility: one column can be used for multicomponent separations and for different mixtures.

Design of batch distillation is discussed, with special attention to thermodynamic properties of the mixture. Selection of process parameters (heat duty, reflux ratio, pressure and cut sizes) plays important role in process optimization. In non-ideal systems the products obtained from batch distillation may be counterintuitive; for example the less-volatile components can be obtained before the more-volatile species.

Typical operating steps depend on the feed mixture and distillation objectives (how many products need to be made and what is the desired purity). These steps may include charging, heating and degassing, establishing reflux, low-boils cut, front cut, pure cut, after cut, pumping out or washing the pot. Operational details of all these steps are discussed; for example the feed line size impacts the time of the batch; or flashing product tanks in the first batch is necessary to reach the purity specification. Most of these operational details are dictated by common sense but they are necessary for running successful batch distillation.

Some examples of plant trials are given.

550022 On the Random Packing Hydraulics for a Corrosive Application

The Chlor-alkali industry uses electrolytic processes to produce chlorine and caustic soda from aqueous sodium chloride solutions. Chlorine drying is one of the stages of chlorine processing which removes moisture from chlorine by using sulfuric acid. Two or more columns are employed in series for chlorine drying by counter-current contact of gas flow with acid flow in a packed column. For the latter, material selection is pivotal to the process design because even small traces of water in chlorine can be highly corrosive. Water content in chlorine and sulfuric acid drive the material selection, which in turn is dependent on each material's chemical resistance, temperature rating, mechanical strength and cost. Design and economic considerations that lead to selecting materials for the column shell, internals and high capacity random packing are discussed in this paper. Quantitative comparisons of the different packing materials in terms of their hydraulic properties will be shown. This will be supported using select commercial application examples.

550226 Necessity of Curvature in Distillation Boundary for Feasible Extractive Distillation Systems in Separating Maximum-Boiling Azeotropes

Extractive distillation is commonly used in industry for separation of azeotropic mixture. A heavy entrainer is added into an extractive distillation column to greatly enhance the relative volatility of one component so that it can go to the distillate of this column while the entrainer and the other component go to the column bottoms. These two components can easily be separated in an entrainer recovery column with heavy entrainer at bottom stream for recycling to the first extractive distillation column. For the separation of maximum-boiling azeotropes, design configuration of the two-column system is exactly the same as the ones for minimum-boiling azeotropic systems. However, because of different topology of ternary diagram for the maximum-boiling system, unique design consideration should be taken to properly design such separation system. Unlike minimum-boiling system having no distillation boundary in the ternary system, there will be a distillation boundary for the maximum-boiling system with a heavy entrainer. Depending on the degree of curvature of distillation boundary, feasible minimum entranier-to-feed ratio can be estimated with given product purity specifications. This minimum ratio can be considered as another screening tool for selecting suitable entrainer of maximum-boiling azeotropic systems. Two examples are presented to demonstrate the important findings for the separation of maximum-boiling azeotrope via extractive distillation system. In the first example of separating acetone and chloroform, over 20% savings in total annual cost and energy can be realized by using a newly proposed entrainer of N-methyl-2-pyrrolidone for this system as opposed to using literature suggested dimethyl sulfoxide or ethylene glycol as entrainer. In the second example of separating phenol and cyclohexanone, the savings in total annual cost and energy are even more significant by using a newly proposed entrainer for this system as opposed to using literature suggested acetophenone adding into this separation system.

Neural Network based Soft Sensor for Pilot-Plant Distillation Column

Thaer A. Abdulla[1], Mohanad Osman[2] and Thomas F. Edgar[1]
(1)The University of Texas at Austin, Austin, TX, US, (2)Canadian Natural Resources Limited, Calgary, AB, Canada

Abstract

The development of an inferential soft sensor for a pilot-plant distillation column separating an ethanol-water mixture using neural network (NN) method has been investigated in this work. Inferential sensors are increasingly used in the process industries to infer the value of the main quality variable while utilizing much easier to measure secondary variables of the process. The lags between the input variables and the output variable vary due to changes in operating conditions. Previous studies have introduced different methods to estimate lags for input and output variables, but all of them have assumed these lags to be constant regardless of the changes in the operating conditions. In this work, an inferential sensor that can infer the composition of ethanol at the top product using time lags for the input variables and varied first-order time constant lags with the output variable is developed. The developed inferential sensor is based on a neural network (NN) model. This NN model is built in MATLAB/Simulink. An OPC link is established in order to communicate between MATLAB/Simulink and the Emerson DetlaV Distributed Control System (DCS) of the pilot-plant distillation column. Principal Component Analysis (PCA) and Projection to Latent Structures (PLS) methods are used in this work to remove the outliers from the input variables set and to determine the most correlated values of the input variables and their lags with the output variable Xa (ethanol mole fraction of distillate product) respectively. The model adaptively selects the correct first-order time constant lags of an output variable according to the instantaneous operating condition (the composition of ethanol is increased or decreased) and assigns a best value for each case. The prediction performance of the proposed NN models (with and without time lags for input variables) is illustrated using experimental data from the pilot-scale ethanol-water distillation column. More than 400 samples were collected to create and validate the results of NN models. The proposed NN model with time lags for input variables and varied first-order time constant lags for output variable gave lower error (RSME=0.01) compared with the NN model without any time lag for input and output variables (RSME=0.06). In summary, a high accuracy soft sensor for the ethanol composition of the top distillation product has been developed and validated. Based on this soft sensor, an inferential PI controller and Model Predictive Controller (MPC) for this pilot-plant column will be developed in future work.

I. Introduction

Recent advancements in measurement techniques in the process industries and increasing plant data collection and storage capabilities have encouraged the trend to estimate some important but difficult to measure process variables from the data collected of other easy to measure variables. The mathematical models used for this estimation are called Soft Sensors or Inferential Sensors. Inferential sensors can be classified into first principles model (FPM) inferential sensors and data-driven inferential sensors. An FPM inferential sensor is constructed by rigorous mathematical derivation of the model from the first principles chemical and physical equations that describe the process. Example of FPM inferential sensors can be found in [1]. FPM inferential sensors suffer from two main drawbacks: high computational cost due to complexity and model inaccuracies due to the simplification assumptions required to solve the first principles equations. Consequently data-driven inferential sensors which are constructed based purely on the process data without the need of knowledge of the process fundamentals are gaining more attention. A number of hybrid inferential sensors based on combinations of the two types have also been introduced [2, 3, 4]. Generally, these hybrid inferential sensors use a data-driven regression technique to estimate some of the parameters in the first principles equations that are two complicated to be driven mathematically. A comprehensive review on inferential sensors can be found in [5, 6].

Generally, data-driven inferential sensors are developed through the following steps: (1) Data pre-processing, (2) Selection of model building technique, (3) Selection of important process variables and their lags, and (4) Model building/validation. The first three steps will be explained in more detail below.

Data pre-processing step mainly includes removal of outliers and data normalization. Sensor faults, signal noise, human errors and unusual operating conditions can cause variables to have unjustifiable extreme values in some observations, these observations are labeled outliers [7]. Including outliers during the construction of the inferential sensor can deteriorate its performance; hence their removal from the data set is essential. In some cases, outliers can be identified via human practical knowledge of the process. Data during unrepresentative operation periods (e.g., start up, shut down) or considered to by the plant experts to be having unusual values based on engineering judgment are labeled outliers and excluded. A more common practice is to identify outliers by the help of statistical analysis of the data {median absolute difference (MAD) [8], principal component analysis (PCA) [9], and Hotelling's T^2 statistics [10]}. Different plant data variables usually have significantly different ranges of their numerical values and hence significantly different variances. For the variables with higher variance to not have disproportionate influence in the model construction data, normalization (or scaling) is crucial. There are several methods of normalization: min-max normalization, unit-variance normalization and zero-mean unit-variance normalization [9].

Selection of model construction technique will depend on many factors; some of them are the degree of process nonlinearity, level of variables collinearity, the importance of the multivariate aspect of the process, number of process variables, the amount of the collected data, level of noise, degree of knowledge of process fundamentals and dynamics and the required state of tradeoff between model accuracy and complexity. The most common techniques used in the construction of inferential sensors are multivariate analysis techniques (PCA and PLS [11, 12], NPCA and NPLS [13, 14], and DPLS [9]), neuro-fuzzy (NF) [15, 16], and artificial neural networks (ANN) [17, 18, 19]. The development of ANN and its applications in chemical engineering can be seen in [20]. The performance of these models can be measured by several indexes and one of the most common indices is the root square mean error (RSME). The model with lower RSME will have the highest accuracy.

Finally, selection of input variables and their lags is very important in the soft sensor development process. Using redundant and collinear input variables increases the complexity of the model and has negative effects on the performance of the inferential sensor. Hence, selecting only a subset of the input variables that have the least co-linearity and the most correlation with the output is an important step of developing an inferential sensor [21]. Typically, the choice of the important variables and their corresponding lags is done simultaneously. The choice of the correct delays for the input variables has a high effect on the quality of the inferential sensor [22]. If the correlation between different delayed versions of an input with the output is studied, the input version with the correct delay should have the highest correlation value with the output. Based on this theory the linear correlation between each of the inputs in different models and the output is used to identify the correct lag for each input variable in [23]. One of the problems of this method is that when the value of the correlation of each input variable with the output variable is calculated individually, it ignores the effect of the other input variables. To solve this problem, Hoskins and Himmelblau [17] have suggested the use of the partial correlation as the identification criteria of the correct lag. Both methods can detect only the linear correlation and fail to work if the relations between the inputs and the output are nonlinear. A method that is applicable for both linear and nonlinear models was proposed in [24] based on Mallows' statistics that use a standardized total square error as a criterion for the selection of the inputs. Another method for nonlinear models, based on the computation of Lipschitz quotient was proposed in [25]. The last two methods suffer from the problem of high computational load. A much simpler method based on partial least squares (PLS) regression technique was proposed in [26].

II. METHODOLOGY

A- ANN Development

Two ANN models are trained for the prediction of ethanol mole fraction of the distillate product (Xa), first model has input lags and the second without input lags. Both NN models are Feed Forward (FF) networks. Levenberg-Marquardt back-propagation with momentum technique is used as the training algorithm for both networks. Trial and error approach is used to find the optimum number of hidden layers, number of neurons of each hidden layer and the activation functions. The network structure of both models (as shown in Table 1) that has the minimum RMSE on the test set is selected for the inferential sensor. Based on this, the structure of four layers (three hidden layers and an output layer) with the neurons (5, 3, 3) and (10, 7, 3) for the network with input lags and the one without input lags respectively has the best performance (RSME=0.02 and 0.04 for networks with input lags and without lags respectively). The activation function of the first hidden layer of both networks was Log-Sigmoid activation function (logsig) and pure-linear activation function (purelin) for other layers of both networks. The description of these networks can be seen in Figures 4 and 5 and Table 1.

In this work, zero-mean unit-variance normalization method has been used for data normalization and PCA method is used for removing the outliers from the input variables set as shown in Figure 1. A PLS method was used to determine the most correlated values of the input variables with the output variable (ethanol mole fraction) (as shown in Figure 2) and the most correlated values of the input lags with the output variable (ethanol mole fraction) (as shown in Figure 3). The variable influence plot (VIP) represents a weighted sum of squares of the weights (WSSW) assigned to the variables in the PLS model, which summarizes the influence and thus the importance of each variable in the prediction of the primary output. Figures 2 and 3 show the VIP of the variables and their lagged versions respectively. The highest WSSW value for the input variables with their lags is selected. For each of the selected input variables, a sequence of 5 lags (k-2, k-4, k-6, k-8, k-10) was obtained with two sample delay between each other, then the variables with their lagged versions were used in the PLS model. Based on these criteria, the following input variables with their lags were selected [4 min lags for top vapor temperature (T200), tray 2 temperature (T102), and tray 3 temperature (T103) were selected while 2 min lag for top column pressure (P081) was selected].

B- Equipment and Experimental Work

The equipment studied in this work is a pilot-scale distillation column of ethanol-water system. The process flow diagram (PFD) of this column can be seen in Figure 6. The column is capable to work either in batch or continuous operations. It has multi input ports to each tray in addition to the main feed entrance to the horizontal reboiler. There are 2 windows at the two bottom trays to watch the column behavior during the operation. The column is 6-inch inner diameter with 10-inch tray spacing, consisting of 10 sieve trays, condenser, reflux drum, and reboiler, and works under atmospheric pressure. The horizontal reboiler is initially filled with feed (20-22 gallon of 20-25 wt% of ethanol). Tables 2 and 3 show the design specifications and the nominal operating conditions of the column. It is desired to develop an inferential sensor that can accurately predict the mole fraction of ethanol (XA) at the top distillate product (the output) using different input variables of the process. The DCS control system (DeltaV) is used for controlling the column operation. The descriptions of the process variables tags can be seen in Table 6.

The column pressure is controlled using nitrogen. There are two valves, one for nitrogen supply and the second for the vents of the column. The PID pressure controller determines which valve will be opened. The subcooled condensate temperature (45 °C) is controlled by the water-cooling flowrate. The reflux drum level is controlled by the flow rates of reflux and distillate streams. Reboiler steam, feed, product, and reflux flow rates are controlled using PID controllers. The reflux drum level controller is usually at Auto mode (Set Point=10 inch) while either the reflux flow controller will be in cascade mode (constant distillate flow operation, D=0.35 l/min) or distillate flow controller will be in cascade mode (constant reflux flow operation, R=0.75 l/min) or both of them will be in cascade mode (constant reflux ratio, R=3).

For the development of the inferential sensor, data were generated using experimental tests. The column was operated at different operating conditions [feed amount (70-85 liter) and feed composition (0.2-0.3 mass fraction of ethanol)] as shown in Table 4. A series of step changes in the reflux ratio (0.5-5) and reboiler steam flow rate (0.55-0.95 kg/min) are introduced as excitation signals to produce a wide range of temperature profiles and top product concentrations. Samples from the distillate product stream are collected regularly at fixed interval (5 to 6 min). The process (input/output) variables are stored within DeltaV (DeltaV history collection) and can be exported to excel sheet at a sampling interval (10 sec or more). Around 400 samples were collected with their corresponding operating conditions. These samples are segmented into two subsets: 160 samples training and testing sets and 240 samples validating set. The RSME of both networks during the validation step was (0.03 for model with input lags and 0.06 for model without input lags, see Figure 7).

C- Output Delay Implementation

Most of the previous works of NN application for distillation columns {estimation and simulation applications [27, 28], optimization applications [29, 30], and control applications [31, 32]) did not account for the whole dynamic path (dynamic behavior) of the output variables (product compositions) and focused only on the final steady state of the output variables so they either did not use any time constant delay for the outputs or just used a constant delay for the outputs. In order to improve the accuracy of our soft sensor during the whole dynamic path of the ethanol mole fraction (output variable) so it will be useful for the inferential control purpose, a variable (updated) time constant delay for the output is used in addition to the constant lags for the input variables. From our experimental investigations, it takes more time (slow response) when we want to increase the composition of ethanol (in other words, long dynamic path of the ethanol composition until reaching the final steady state from below) while it takes less time (fast response) when we want to decrease the ethanol composition (in other words, short dynamic path of the ethanol composition until reaching the final steady state from above). So, several pairs of first-order time constant lags are used for both cases (ethanol composition is increased/deceased). A past value for the ethanol composition (0.5 min earlier) is subtracted from the current value of ethanol composition. If the result of the subtraction is positive (ethanol composition is increased), those first-order time constant lags (3 min, 5min, 7min, and 9min) are tried and if the result of the subtraction is negative (ethanol composition is going down), those first-order time constant lags (1min, 3 min, and 5min) are tried. So, the following pairs of first-order time constant lags {(1min, 3min), (1min, 5min), (1min, 7min), (3min, 3min), (3min, 5min), (3min, 7min), (5min, 5min), (5min, 7min), (5min, 9min); note the first value of these pairs is when the composition is going down while the second value is when the composition is going up} are used to create different sensors (AI91E, AI91EPV1 , AI91EPV2, AI91EPV3, AI91EPV4, AI91EPV5, AI91EPV6, AI91EPV7, AI91EPV8) as shown in Table 5 .

III. RESULTS & DISCUSSION

A- Initial Validation Results of the Soft Sensor

Figure 7 shows a comparison between the performance of the ANN inferential sensor model with input lags and the ANN inferential sensor model without input lags for the online prediction based on the validation set. It is clear that the inferential sensor model with input lags has significantly outperformed the one without lags, achieving a considerably lower RMSE of (0.03).

B- Improved Validation Results of the Soft Sensor

Online tests were conducted for the proposed adaptive selection of time constant lags inferential sensor. It is clear that the varying output lags inferential sensor have surpassed the ones with only constant input lags or without input lags in the prediction accuracy in all these tests as shown in Table 5 and Figure 8. The sensor with best time constant lag configuration (3min, 7min) for the output variable (ethanol mole fraction) in addition to the constant input lags of the model has been selected (AI91EPV5) in our work so the accuracy of this sensor has been increased (RSME=0.01).

IV. CONCLUSIONS AND DIRECTIONS FOR FUTURE WORK

In this work, it was found that the NN model with input variable lags has more accuracy compared with the NN model without input variable lags. The time constant lags of the output variable (ethanol mole fraction of top product) are significantly dependent on the instantaneous value of the operating conditions (whether the ethanol composition is increased or decreased). The NN model with the updated time constant lags for output variable in addition to the input variables lags developed in this work was able to account for the whole dynamic path (dynamic behavior) of the output variables (product composition). This model is found to outperform a model with the input variables lags or the one without input variables lags. The high accuracy of the inferential sensor in tracking the value of the actual top product composition closely made it possible for an effective control scheme to be applied on the system. Based on this inferential sensor, inferential PI controller and MPC (DeltaV MPC/Aspen DMCplus) will be developed and their performance will be compared for future work.

V. REFERENCES

[1] Luo, R.F., Shao H.H., Zhang Z.J. (1995), Fuzzy-Neural-Net-Based Inferential Control for A High-Purity Distillation Column, Control Eng. Practice, Vol. 3, No. 1, pp. 31-40

[2] Gnoth, S., Jenzsch, M., Simutis, R., Lübbert, A. (2008), Product formation kinetics in genetically modified E. coli bacteria: inclusion body formation, Bioprocess Biosyst Eng, Vol. 31, pp. 41–46

[3] Kim, B. J., Kim, I. K. (2005), An Application of Hybrid Least Squares Support Vector Machine to Environmental Process Modeling1, Parallel and Distributed Computing: Applications and Technologies, Lecture Notes in Computer Science, Volume 3320, pp. 57-85

[4] Fortuna, L., Grazaini, S., Rizzo, A., Xibilia, M. G. (2007). Soft Sensors for Monitoring and Control of Industrial Processes. Springer-Verlag, London.

[5] Kadleca, P., Gabrys, B., Strandtb, S. (2009), Data-Driven Soft Sensors in The Process Industry, Computers and Chemical Engineering, Vol. 33, No. 4, pp. 795-814

[6] Lin, B., Recke, B., Knudsen, J. K. H., Jørgensen, S. B. (2007), A systematic approach for soft sensor development, Computers and Chemical Engineering, Vol. 31, pp. 419–425

[7] Shu Xu, Bo Lu, Michael Baldea, Thomas F. Edgar, Willy Wojsznis, Terrence Blevins, Mark Nixon, Data cleaning in the process industries, Rev Chem Eng 2015; 31(5): 453–490

[8] Liu, J., Min, K., Flan C., Chang, K. S. (2000), Robust Nonlinear PLS Based on Neural Networks and Application to

Composition Estimator for High-Purity Distillation Columns, Korean J. Chem. Eng., Vol. 17, No. 2, pp. 184-192

[9] Eriksson, L., Johansson, E., Kettaneh-Wold, N., and Wold, S. (2001). Multi- and Megavariate Data Analysis. Umetrics Academy, Sweden.

[10] Shin, J., Park, S. (2000), Design of a Composition Estimator for Inferential Control of High-Purity Distillation Columns, Chemical Engineering Communications, Vol. 178, No. 1, pp. 221 - 248

[11] Fujiwara, K., Kano, M., Hasebe, S. (2008) Correlation-Based Just-In-Time Modeling for Soft-Sensor Design, Computer Aided Chemical Engineering, Vol. 25, pp. 471-476

[12] Zamprogna, E., Barolo, M., Seborg, D. E. (2005). Optimal selection of soft sensor inputs for batch distillation columns using principal component analysis. Journal of Process Control, Vol. 15, pp. 39-52.

[13] Mark A. Kramer, Nonlinear Principal Component Analysis Using Autoassociative Neural Networks, AICHE Journal, Vol. 37, No. 2, February 1991

[14] E. C. Malthouse, A. C. Tamhane and R. S. H. Mah, Nonlinear partial least squares, Computers chem. Engng Vol. 21, No. 8. pp. 875-890, 1997

[15] Arauzo-Bravo, M. J., Cano-Izquierdo, J. M., Gomez-Sanchez, E., Lopez-Nieto , M. J., Dimitriadis, Y. A., Lopez-Coronado, J. (2004), Automatization of a penicillin production process with soft sensors and an adaptive controller based on neuro-fuzzy systems, Control Eng. Pract., vol. 12, pp. 1073–1090

[16] de Canetea, J. F., del Saz-Orozcoa, P., Gonzalez-Pereza, S. (2008), Application of Adaptive Neurofuzzy Control Using Soft Sensors to Continuous Distillation, Computer Aided Chemical Engineering, Vol. 25, pp. 465-470

[17] Hoskins, J. C., Himmelblau, D. M. (1988), Artificial Neural Network Models of Knowledge Representation in Chemical Engineering, Comput. Chem. Eng., Vol. 12, No. 9/10, pp. 881-890

[18] Gonzaga, J.C.B., Meleiro, L.A.C., Kiang, C., Filho, R. M. (2009), ANN-based soft-sensor for real-time process monitoring and control of an industrial polymerization process, Computers and Chemical Engineering, Vol. 33, No. 1, pp. 43–49

[19] Wang, X., Ding, Y., Shao, H. (1998), the Improved Radial Basis Function Neural Network and its Application, Artificial Life and Robotics, Vol. 2, No. 1, pp. 8-11

[20] Venkat Venkatasubramanian (2019), The Promise of Artificial Intelligence inChemical Engineering: Is It Here, Finally?, AIChE Journal, Vol. 65, No. 2, pp. 466-478

[21] Feng, R., Shen, W., Shao, H. (2003), A Soft Sensor Modeling Approach Using Support Vector Machines, Proceedings of the 2003 American Control Conference, Denver, Colorado, June 4-6, Vol. 5, pp. 3702-3707

[22] Lin, B., Recke, B., Knudsen, J. K.H., Jørgensen, S. B. (2007), a Systematic Approach for Soft Sensor Development, Computers and Chemical Engineering, Vol. 31, No. 5, pp. 419-425

[23] Zamprogna, E., Barolo, M., Seborg, D. E. (2005), Optimal Selection of Soft Sensor Inputs for Batch Distillation Columns Using Principal Component Analysis, Journal of Process Control, Vol. 15, No. 1, pp. 39-52.

[24] Gomm, J. B., Evans, J. T., Williams, D. (1997), Development and Performance of a Neural-Network Predictive Controller, Control Engineering Practice, Vol. 5, No. 1, pp. 49-59

[25] Fortuna, L., Graziani, S., Xibilia, M. G., Napoli, G. (2005), Soft sensors for product quality monitoring in debutanizer distillation column, Control Engineering Practice, Vol. 13, pp. 499-508

[26] Warne, K., Prasad, G., Rezvani, S., Maguire, L. (2004), Statistical and Computational Intelligence Techniques for Inferential Model Development: a Comparative Evaluation and Novel Proposition for Fusion, Engineering Applications of Artificial Intelligence, Vol. 17, No. 8, pp. 871- 885

[27] Vijander Singh, Indra Gupta, H.O. Gupta (2007), ANN-based estimator for distillation using Levenberg–Marquardt approach, Engineering Applications of Artificial Intelligence, Vol. 20, pp. 249–259

[28] M.T. Vafaei, R. Eslamloueyan, Sh. Ayatollahi (2009), Simulation of steam distillation process using neural networks, chemical engineering research and design, Vol. 8 7, pp. 997–1002

[29] Funmilayo N. Osuolale, Jie Zhang (2016), Energy efficiency optimization for distillation column using artificial neural network models, Energy, Vol. 106, pp. 562-578

[30] Funmilayo N. Osuolalea, Jie Zhanga (2017), Thermodynamic optimization of atmospheric distillation unit, Computers and Chemical Engineering, Vol. 103, pp. 201–209

[31] P. Dutta, R. Russell Rhinehart (1999), Application of neural network control to distillation and an experimental comparison with other advanced controllers, ISA Transactions, Vol. 38, pp. 251-278

[32] Neha Sharma, Kailash Singh (2012), Model predictive control and neural network predictive control of TAME reactive distillation column, Chemical Engineering and Processing, Vol. 59, pp. 9– 2

VI. APPENDICES

Appendix (A): Figures

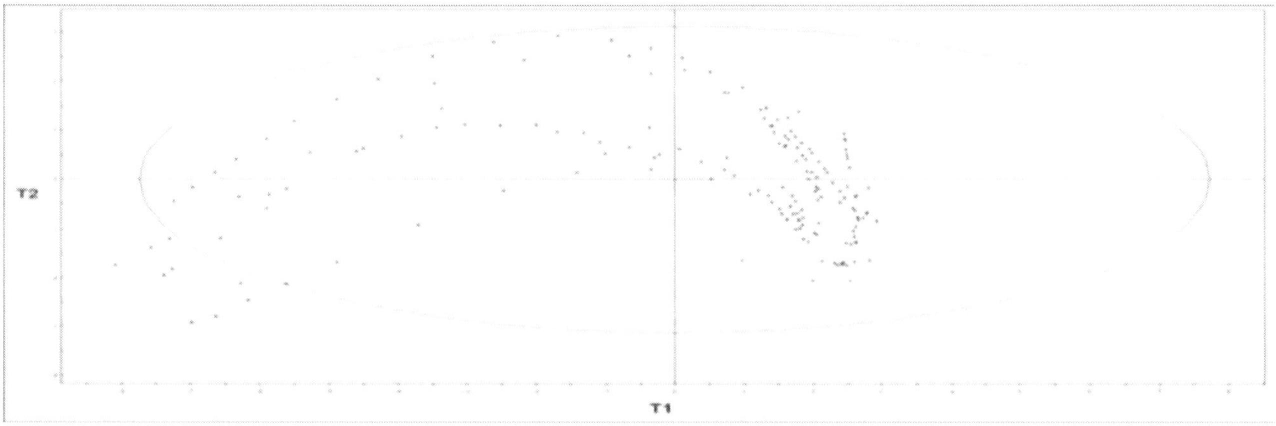

Figure 1 Score plot of the first two principal components of a PCA study for 169 observations

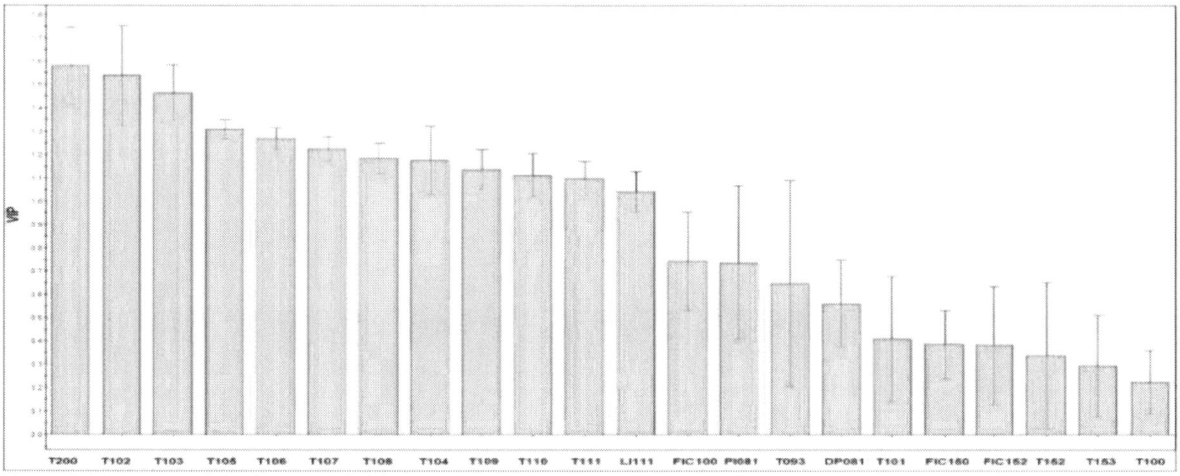

Figure 2 PLS-VIP of the input variables

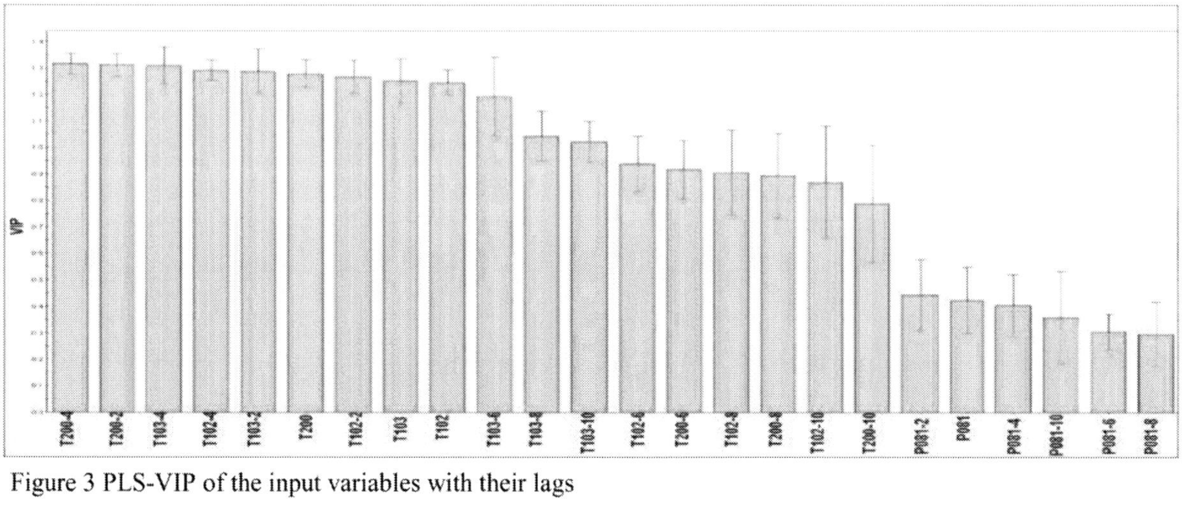

Figure 3 PLS-VIP of the input variables with their lags

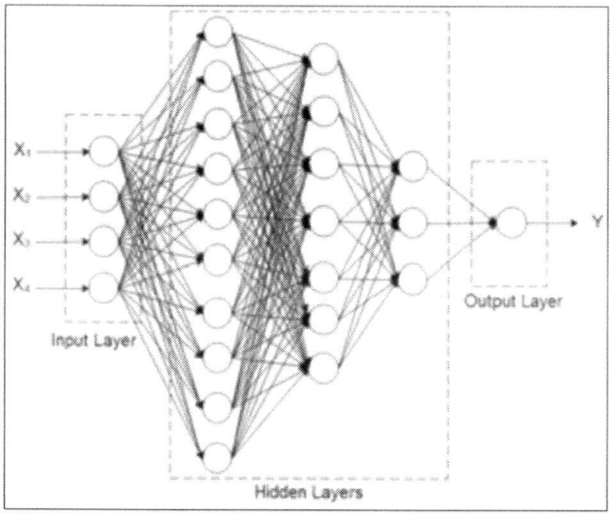

Figure 4 Structure of the NN Model without Input Delays

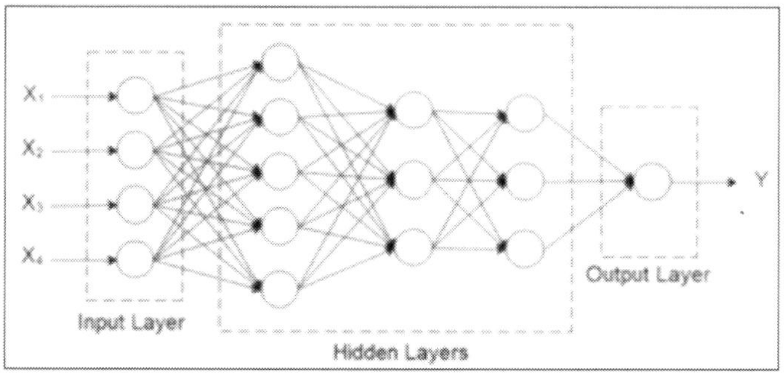

Figure 5 Structure of the NN Model with Input Delays

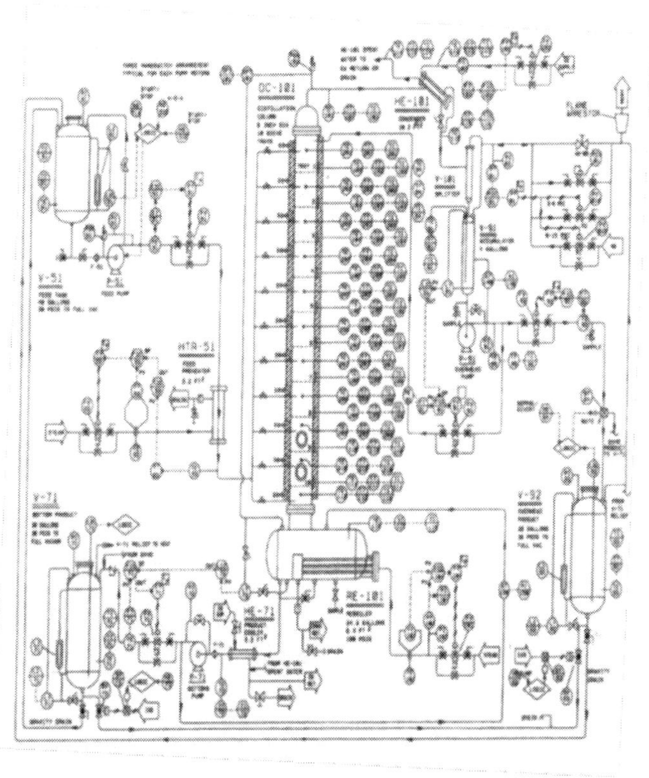

Figure 6 PFD of Ethanol-Water Distillation Column

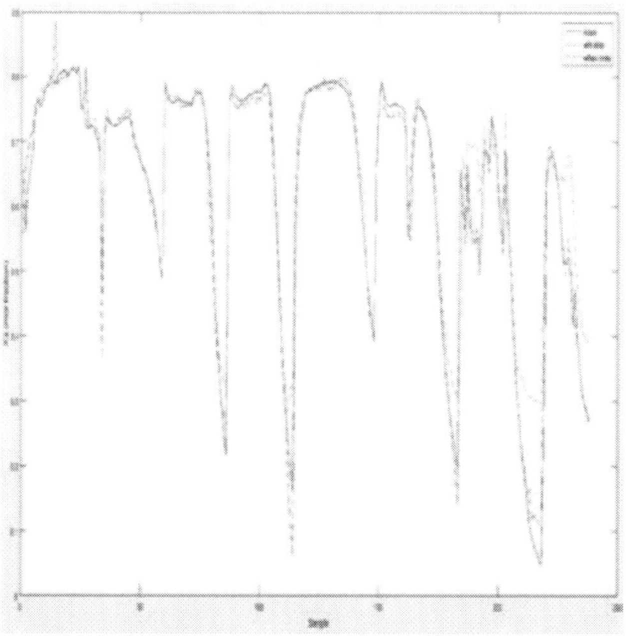

Figure 7 NN Soft Sensor Initial Validation Results

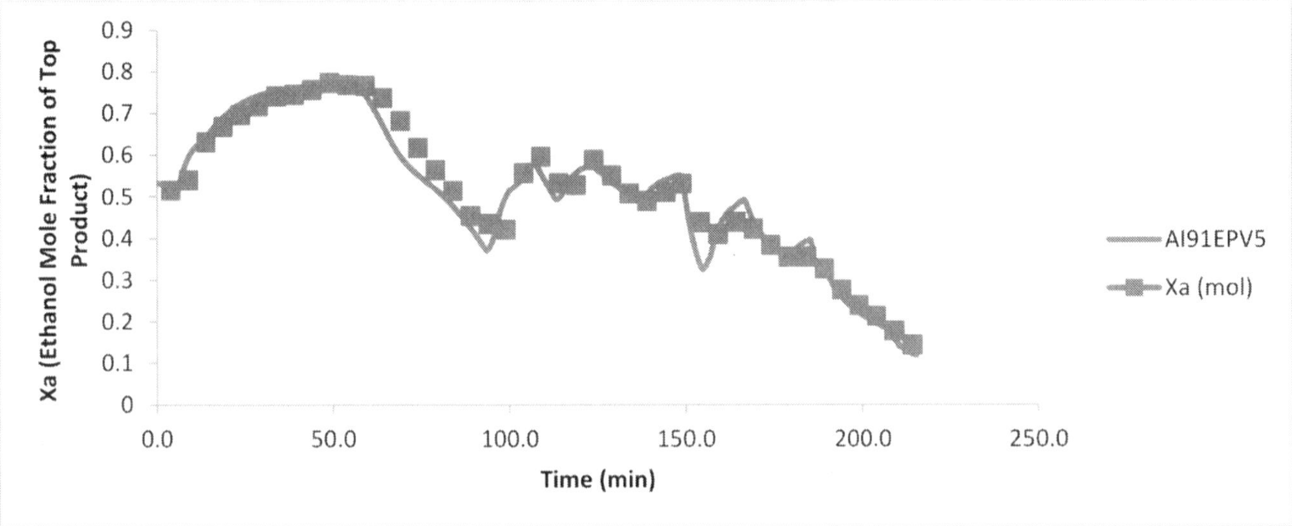

Figure 8 Improved NN Soft Sensor Validation Results

Appendix (B): Tables

Table 1 Neural Networks Structure

Name	**Type**	**Training Algorithm**	**No. of hidden Layers**	**No. of Neurons in each hidden layer**	**Activation Function in the hidden and output layers**
NN without input lags	Feed Forward	Levenberg-Marquardt back-propagation with momentum technique	3	10, 7, 3	Logsig, Purelin, Purelin, Purelin
NN with input lags	Feed Forward	Levenberg-Marquardt back-propagation with momentum technique	3	5, 3, 3	Logsig, Purelin, Purelin, Purelin

Table 2 Ethanol-Water Batch Distillation Column Specifications

Sieve trays specifications	
Number of sieve trays	10
Diameter (inch)	6.065
Tray space (inch)	10
Weir height (inch)	0.5
Lw/D	0.725
Fractional Active Area (Aa/A)	0.7
Fractional Hole Area (Ah/Aa)	0.29
Discharge Coefficient	0.8
Sieve Tray Hole Diameter (inch)	0.03125
Downcomer Area Fraction (Ac/A)	0.0278
Reboiler Pot Specifications	
Geometry	Horizontal Cylinder with Elliptical Heads
Length (inch)	14
Diameter (inch)	33
Heating Transfer Medium	Steam

Contd.	
Steam pressure (bar)	2
Steam Mass Rate (Kg/min)	0.8 kg/min
Steam Coil Configuration	
Bottom height (inch)	0
Top height (inch)	4.25
Area (m2)	0.596
Condenser Specifications	
Condenser Type	Shell and tube (2 pass design, city water on tubes, process condensate on shell side), Total Condensation with subcooling
Subcooled temperature (°C)	45
Condenser exchange area (m2)	1
Reflux Drum Specifications	
Geometry	Vertical cylinder with elliptical heads
Height (inch)	14
Diameter (inch)	6.5
Product tank Specification	
Geometry	Vertical cylinder with elliptical heads
Height (inch)	36
Diameter (inch)	14

Table 3 Water-Ethanol Batch Distillation Nominal Operation Conditions

Operation Conditions	
Feed Initial Charge (Gal or Lit)	20-22 Gal or 75-85 Lit
Feed Initial Composition (%wt)	18-28 %wt ethanol
Initial charge conditions (Temp and Pressure)	25 °C and 101 Kpa
Reflux Ratio, Constant Distillate, OR Constant Xd	3, 0.35l/min, Or 91%wt ethanol
Stopping criteria	Final ethanol mass fraction at Pot (Xw=0.005)
Top pressure (kpa)	101 Kpa
Reboiler Duty	0.8 kg/min sat. steam @3 bar

Table 4 Range of Operating Conditions used for the Training of NN

Operating parameter	**Range**
Reboiler steam flow rate (kg/min)	0.55 - 0.95
Distillate flow rate (l/min)	0.20 - 0.80
Reflux flow rate (l/min)	0.20 - 0.80
Reflux Ratio	0.50 - 50
Initial Feed Amount (Gallon)	17.50 - 22.50
Initial Ethanol Mass Fraction at Feed	0.18 - 0.28

Table 5 Soft Sensors Performance

Sensor	Model Type	RSME
AI91	NN without Input Lags	0.04013
AI91B	NN with Input Lags	0.02239
AI91E	NN with input lags and output time constants (1min, 3min)	0.01898
AI91EPV1	NN with input lags and output time constants (1min, 5min)	0.01897
AI91EPV2	NN with input lags and output time constants (1min, 7min)	0.01897
AI91EPV3	NN with input lags and output time constants (3min, 3min)	0.01746
AI91EPV4	NN with input lags and output time constants (3min, 5min)	0.01750
AI91EPV5	**NN with input lags and output time constants (3min, 7min)**	**0.01710**
AI91EPV6	NN with input lags and output time constants (5min, 5min)	0.02280
AI91EPV7	NN with input lags and output time constants (5min, 7min)	0.02313
AI91EPV8	NN with input lags and output time constants (5min, 9min)	0.02334

Table 6 Process Variables Tags Descriptions

Tags	Descriptions
T200	top column vapor temperature
T093	Subcooled condensate liquid temperature (°C)
T100	reflux temperature
T101	tray 1 (top tray) liquid temperature (°C)
T102	tray 2 liquid temperature (°C)
T103	tray 3 liquid temperature (°C)
T104	tray 4 liquid temperature (°C)
T105	tray 5 liquid temperature (°C)
T106	tray 6 liquid temperature (°C)
T107	tray 7 liquid temperature (°C)
T108	tray 8 liquid temperature (°C)
T109	tray 9 liquid temperature (°C)
T110	tray 10 (bottom tray) liquid temperature (°C)
T111	reboiler liquid temperature (°C)
LI111	reboiler level (cm)
T152	inlet water cooling temperature (°C)
T153	outlet water cooling temperature (°C)
FIC152	water cooling flowrate (l/min)
FIC100	reflux flowrate (l/min)
FIC150	steam mass flowrate (kg/min)
DP081	column differential pressure (kpa)
PI081	top column pressure (kpa)

548001 Mass Balance and Loss Control, a Study in the Process to Identify and Quantify Losses at a Refinery and Best Practices for Managing These Losses

Hydrocarbon loss reduction programs can save $2-10 million per year for an average sized refinery. This can be not only from the value of loss material but also introduction of errors in yield models that affect overall optimization.

Knowing where to start in implementing a loss control program though can seem like an overwhelming task. This presentation will take you through the process of evaluating losses including identifying the financial impact in terms of value and risk of occurrence to justify reducing losses by improvements in measurement certainty.

Losses are calculated through the material and energy balances of the plant. Best practices for both static and dynamic measurement will be covered. We will discuss the most common sources of error in calculating these balances.

Losses can be both controllable and uncontrollable, and are both real and apparent (due to measurement problems). Controllable losses include flares, plant effluent, tank evaporation and solid waste, all of which must be minimized. Before steps can be taken to reduce these losses however, the real and apparent losses need to be well understood.

550877 Improve Refinery Profitability with Modus Technology

Improve Refinery Profitability with Modus™ Technology

Catalytic Cracking (FCC) and Coking are the two key conversion processes in a petroleum refinery. Each technology yields light olefins, e.g. ethylene, propylene, butylenes as a byproduct. FCC and Coking Units include a gas recovery section where light olefins and gasoline/naphtha are separated. Propylene/Butylenes are normally processed in alkylation or oligomerization facilities to produce additional high octane gasoline. Unrecovered light olefins, including all of the ethylene are normally routed to the refinery fuel system. As facilities age and operate at higher capacities the recovery of propylene/butylenes diminishes with increased quantities of these valuable compounds burned as fuel. Some refiners opt to modify their gas recovery units, sometimes including cryogenic processing to increase propylene/butylenes recovery. None of these routes capture the value of ethylene for the refiner. The quantity of light olefins burned in the fuel system can become even more significant as FCC catalysts are employed that promote higher light olefin yields.

Wood has partnered with Siluria Technologies to commercialize Modus technology which converts light olefins contained in FCC/Coker fuel gas product streams to a high octane gasoline blending component. Modus is a modularized, drop in solution that requires minimal interruption of plant operations for implementation. The reaction pathway produces a high octane gasoline product containing a mixture of aromatics, olefins and paraffin isomers.

This presentation will describe the Modus technology and its application in a number of processing scenarios.

Applying Modeling/Simulation to Deliver Insight, Understanding, and Economic Value

Authors: P. Braswell and S. Tieri

Business and industry are increasingly challenged to develop new and improved technologies & processes that deliver increased customer value while maintaining or improving on commercial economic targets. The process insight and understandings obtained though models are essential to informed business strategy, targeted innovation, and tactical operational decisions. Process models are critical to consistently safe and successful implementation of applied science and engineering to deliver innovative products, solutions, and value to our customers. DuPont applies a diverse suite of modeling & process simulation tools to enable informed comparisons and decisions about optimal process configuration, alternative operations, detailed equipment design, and inherently safer operating conditions.

The included case studies are from typical process analyses used by the businesses to quantify and evaluate process energy use, utility requirements, raw material consumption, and by-product generation and demonstrate the benefit to our organization and accelerate our ability to screen and prioritize technology development opportunities. This paper will illustrate DuPont use of process modeling to identify & evaluate options, optimize energy and material requirements, and justify technology implementation and/or process upgrades to deliver economically viable & sustainable technology to the marketplace.

550613 Tracking the Impact of Operating Condition Changes across Refinery Units

Given the degree of interconnection between different refinery processes, a viable way to improve product quality and refinery conversion unit efficiency is to adjust process parameters of separation processes, petrochemical plants, or other refinery conversion units that feed a given downstream conversion process.

This work shows the simulation of different types of refinery processes to study the impact of changes in operating conditions and how the effects propagate across other refinery units. Models were developed with different thermodynamic methods, definition of species set, and rigorous kinetic models to account for the chemical and physical phenomena occurring in each unit.

Three case studies will be presented:

- Changes in straight-run diesel cut point from a crude column and the effect on the hydrotreating unit.
- The impact of recycle H_2S from the amine scrubber on hydrotreater reactor performance and separation.
- The influence of changes in butane isomerization and light paraffin dehydrogenation as feed to the alkylation process.

550761 An Integrated MINLP Model of Simultaneous Crude Procurement Planning and Movement Scheduling for Refineries

The petroleum refining process is very complicated, since huge volumes of crudes should be continuously handled through a highly integrated network with extremely intensive units. As one of the key factors that could significantly affect the refining process, the procurement of feedstock, including detailed types, volume and timing, is still vulnerable to restrict operational requirements of refinery and various uncertainties. Different crudes may possess distinct properties, thus require additional operations or even different processing procedures. On the other hand, the crude procurement cost is comparable to the operating cost in quantity, taking up a significant proportion of the overall cost [1]. And the crude market has become highly volatile since the last two decades [2]. Improper crude procurement plan will inevitably cause serious operating problems and result in economic loss. Thus, both process feasibility and refinery profit margin will be affected. Note that after the crude procurement plan has been made, crude movement schedules, covering from vessels, berths, storage tanks, pipelines, charging tanks to CDU have to be well scheduled to accomplish the smooth and stable crude transfer. In previous studies, these issues were addressed in a sequential manner, generating respective crude purchase plans and crude movement schedules [3,4]. Some works studied a partial combination of crude purchasing and blending operations [5,6]. However, the two crucial elements have not been simultaneously considered in a full scope, which could result in the gap between current solutions and real optimality in terms of both operation feasibility and process economy.

In this study, a new methodology has been proposed to tackle this tough challenge. Both crude procurement and oil movement will be simultaneously considered in order to eliminate the optimality gap that could obstruct the processing efficiency and profitability of refinery. Accordingly, a novel MINLP model has been developed. Through this model, the optimal crude procurement plan and oil transfer schedule can be efficiently determined towards the maximal profit margin. More importantly, the integrated method will ensure that decisions for the two sectors can harmoniously cooperate and eventually benefit the processing of refinery. In addition, realistic requirements have been taken into account in this work, including purchasing constraints, such as crude characteristics and crude delivery time and process requirements, for instance tank inventory and blending specs, so that the model applicability can be guaranteed. The developed model has been tested through various case studies to demonstrate its efficacy and performance.

References

[1] Ji, X., Huang, S., & Grossmann, I., (2015) Integrated operational and financial hedging for risk management in crude oil procurement. Industrial & Engineering Chemistry Research, 54, 9191-9201.

[2] Chen, R., Deng, T., Huang, S., & Qin, R., (2015) Optimal crude oil procurement under fluctuating price in an oil refinery. European Journal of Operational Research, 245, 438-445.

[3] Kallestrup, K., Lynge, L., Akkerman, R., & Oddsdottir, T., (2014) Decision support in hierarchical planning systems: The case of procurement planning in oil refining industries. Decision Support Systems, 68, 49-63.

[4] Zhang, S., & Xu, Q., (2015) Refinery continuous-time crude scheduling with consideration of long-distance pipeline transportation. Computer and Chemical Engineering, 75, 74-94.

[5] Oddsdottir, T., Grunow, M., & Akkerman, R., (2013) Procurement planning in oil refining industries considering blending operations. Computer and Chemical Engineering, 58, 1-13.

[6] Zhang, J., Wen, Y., & Xu, Q., (2012) Simultaneous optimization of crude oil blending and purchase

planning with delivery uncertainty consideration. Industrial & Engineering Chemistry Research, 51, 8453-8464.

550616 Using Big Data Techniques to Tackle Refining Reliability and Optimization

I think we all recognize there is journey to get to Industry 4.0 or Refining 4.0. How do we get to the smart refinery? The smart refinery includes the Internet of Things, Cloud computing, cyber-physical process, and Analytics. Teradata is working with companies on their journey to smart refineries by helping companies with their analytics through IoT and cloud computing at scale of machine learning algorithms. Teradata is providing companies with access to the power of analytics by combining their sensor data and ERP data at granular detail level to gain insights on how to improve costs and increase gross margin in their reliability and optimization processes. We will discuss the methods we apply to to bridge the OT (Operational Technology) and IT(Information Technology) worlds together to have impacts on maintenance programs, energy optimization, and operations. We will also discuss how data integration around applications, data silos , ERP systems have to helping companies not only gain insights but also have financial impact to the bottom line at scale for multiple refineries.

551534 Transforming Abundant Data into Improved Process Reliability

Today, processing facilities are beginning to effectively utilize the abundance of process and asset health data and information as part of an overall IIoT strategy. Although a digital transformation will require investments in technology, one cannot forget the investment in training staff to behave differently (improved) with the new timely information. It should be noted that some jobs will be replaced (e.g. manual data collection), new roles will emerge (e.g. analysts and planners), and other jobs will evolve to be more efficient. Properly trained operations are invaluable in this process, as they not only act on information from analyzed data, but can also make decisions on their own based on sensor data.

Abnormal operation and imminent asset failure can quickly be identified and corrected with properly trained staff to act on this new timely analyzed information. Big data and IIoT will continue to have a role with shortages of skilled resources, thus having the staff be more efficient and effective at their role with additional information and insight into operations.

This presentation will focus on the applications where big data and IIoT have improved process and asset reliability. Although the applications presented will come from oil refineries, the solutions can be applied to other continuous processing facilities.

The Digital Plant

Industrial IoT Thoughts by Timothy Holtan (tpholtan@gmail.com)

The manufacturing manager makes hundred of decisions every week. The decisions range from the current revenue - what products at what volume - to the future expense - how much research and development to stay in front of competition. Excellent decision making combines a lot a data efficiently and objectively at a cost of speed. We will discuss a business process to speed objective decision making.

Manufacturing Manager Challenges

Manufacturing managers such as refinery managers, automobile manufacturing plant managers and nuclear power plant managers make many important decisions every day. How many units to produce. When to add overtime to increase production. What is the best product mix for the plant and for the market.

The manager faces continuous change in the form of technology advancement, regulations, labor availability, competition and other internal and external forces. Manufacturing usually operates on a cycle of first planning, then executing, and then auditing the results. These three stream in parallel as the manager continuously plans the future, executes and present and audits the past.

New Decisions, Legacy Solutions

The best manufacturing manager uses all of the tools available from ERP systems and marketing demand planning solutions and Enterprise Asset Management to make the best decisions to ensure the best future state. Managers use a familiar format to assess their performance. The variance report compares actual results to the plan and measures the difference as success or failure.

The variance report and manufacturing planning unquestionably delivers proven value. The process pushes against the limits of human execution and forecasting. Often the lens of the variance report process exposes the participant's biases. Consider a negative variance in contribution margin as an example of how these biases creep into the planning and execution process. Some managers favor top line growth and others favor cost cutting.

How much extra value could the manufacturing manager gain when he looked beyond personal bias and the staff's collective experience? What if technology offered a cold eye review of every variance and played out what if scenarios 90 days ahead in the face of raw data?

Artificial Intelligence Offers an Alternative

The rise of rapidly scaling artificial intelligence solutions offers the manufacturing manager a solution to gain a third column in the variance reports in addition to actual and plan - the prediction.

New artificial intelligence solutions demonstrate step change capabilities to combine historical data and known future data with both deterministic and nondeterministic relationships to quantify probabilistic outcomes.

Point predictive AI solutions have demonstrated profit improvements in the order of 25% to 100%. Documented examples include GE Digital SmartSignal predictive maintenance solution in the power generation industry and IBM Maximo for maintenance planning. These solutions currently operate within the function of the manufacturing process as a form of human assist and automation. For example, AI based predictive maintenance solutions in the maintenance department offer insights which help shift unplanned maintenance to planned. AI traffic solutions in the logistics department reduce transportation costs. Often a data science team - either a external vendor or an internal function - ensures a successful deployment and ongoing sustain and maintain responsibilities remain in the department.

Historical precedent suggests the AI solutions will gain even more value as the solutions comprehend more and more of the functional data. For example, as process control expanded from single loops to the multivariate advanced control to process unit constraint control value increases. Similarly logistics and production planning expanded from local into plant wide functions and transformed into the economics and planning department and focused their efforts on the mastery of linear programming solutions.

As these solutions grew in value, they specialized into separate functions beyond the original department. The process control department grew out of the operations department to provide a continued focus on utilizing the process control solutions to add value and to ensure advanced process control solutions broke out of the boundaries of the operations department. For example advanced controls could maximize objective functions across upstream and downstream processes which usually operate in data silos.

After the Information Silos

In the digital transformation parlance, this phenomena is known as breaking down information silos. Before economics and value cause AI solutions to break these silos, manufacturing managers will anticipate and accelerate AI adoption and follow the tradition of economics and planning and advanced control and form the Prediction Department

The Prediction Department will organize to use the AI solutions to continuously generate probabilistic data driven predictions for every parameter in the monthly, quarterly and annual variance report.

The manufacturing manager will derive value from the prediction department in three ways.

1. **People and Process**. The Prediction Department solves the problem of what to do with the people as the data silos fail. The manpower associated with maintaining the point solutions and departmental databases dedicated to local Optimizatlon shift to broader coverage to achieve global optimization.

2. **Sustain and Maintain**. Second, the Prediction Department solves both the knowledge transfer and the sustain and maintain challenge of the new AI solutions. Shifting manufacturing expertise to the Prediction Department and cross training the manufacturing incumbents to sustain and maintain the AI solution will allow vendors to handover analytics and predictive functions to a receptive, expert team.
3. **Focused Number Cruncher**. Finally the manufacturing manager will gain the resource to add the prediction column to the variance report.

The Prediction department will generates more comprehensive predictions from data stretching across the organization to include future demand and future costs. AI pushes the human participant to realize opportunities beyond the limits unique history and personal bias.

Vision of Success

How will the manufacturing manager gauge the success of the operations Prediction Department?

When the Prediction Department runs the variance meeting and the focus of the discussion shifts from why the plant produced 10% less profit last month to predicting how the plant could increase profit from 10% to 15% next month, the transformation succeeded.

Sunk cost economics teaches that the future is all that matters and the Prediction Department will focus the AI to change the future.

549968 California Energy Commission Integrated Energy Policy Report

Every two years, the California Energy Commission prepares the Integrated Energy Policy Report (IEPR). The IEPR highlights the implementation of California's innovative policies and the role they have played in establishing a clean energy economy. The 2018 Update report details several key energy issues and will encompass new analyses.

551808 Effects of Elevated CO_2 Concentrations on Yields of Agricultural Crops

That elevated concentrations of CO_2 can increase plant growth has been known since at least 1804, when a scientist named De Saussure reported that peas exposed to high CO_2 concentrations grew better than control plants in ambient air (about 300 ppm at that time). Since then, thousands of experiments with elevated concentrations have been conducted, mostly in greenhouses and other types of enclosures, and almost of them similarly showed increases in growth and/or yield. Indeed, it is standard horticultural practice to enrich greenhouses to about 1000 ppm CO_2 when conditions are cool enough that so that the greenhouses do not have to be ventilated for temperature control. Then, when it was realized that the atmospheric CO_2 concentrations were increasing, scientists wondered what this portended for future plant growth and agricultural productivity in open fields. Therefore, about 30 years ago, free-air CO_2 enrichment (FACE) technology was developed that enabled the air above open-field plots to be enriched with CO_2 for entire growing seasons. Since then, FACE experiments have been conducted on cotton, wheat, ryegrass, clover, potato, grape, rice, barley, sugar beet, soybean, cassava, rape, mustard, coffee (C_3 crops), and sorghum and maize (C_4 crops). Elevated CO_2 (550 ppm from an ambient concentration of about 353 ppm in 1990) decreased evapotranspiration about 10% on average and increased canopy temperatures about 0.7°C. Biomass and yield were increased by FACE in all C_3 species, but not in C_4 species except when water was limiting. Yields of C_3 grain crops were increased on average about 19%, while other C_3 crops were increased even more.

Rate-Based Column Simulation in the Complex Domain

Jaeide B. Jawahill and Ross Taylor
Clarkson University

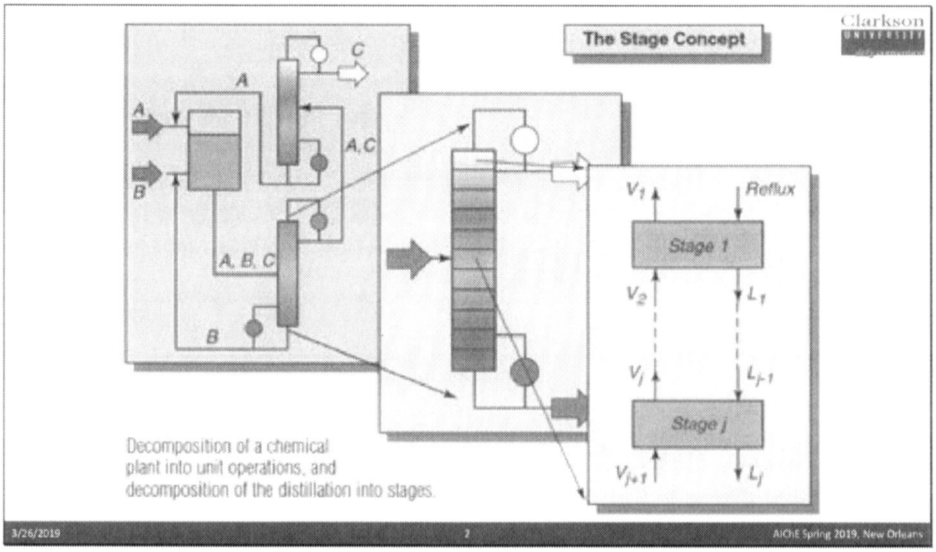

To model a plant like the one shown in here, we decompose the entire plant into smaller units. In this case, the plant contains a distillation column that is shown enlarged in the center panel of the figure. There are many ways to model an entire column, but the most common approach is to divide the column into a number of discrete "stages," as depicted in the third panel. Thus, one question to be addressed is: How do we model these stages?

Chemical engineers have been solving distillation problems by using the *equilibrium-stage model* since 1893 when Sorel outlined the concept to describe the distillation of alcohol. Since that time, it has been used to model a wide variety of distillation-like processes, including simple distillation (single-feed, two-product columns), complex distillation (multiple-feed, multiple-product columns), extractive and azeotropic distillation, petroleum distillation, absorption, liquid-liquid extraction, stripping, and supercritical extraction.

From the late 1950s to the early 1990s, hardly a year passed without the publication of at least one (and usually many more than one) new algorithm for solving these. The evolution of algorithms for solving the stage model equations has been influenced by, among other things, the availability (or lack) of sufficient computer storage and speed, the development of mathematical techniques that can be exploited, the complexity of physical property (K value and enthalpy) correlations, and the form of the model equations being solved. Many computer-based methods are discussed at length in a number of textbooks [see, e.g., Holland, *Fundamentals of Multicomponent Distillation,* McGraw-Hill, New York, 1981 King, *Separation Processes,* 2d ed., McGraw-Hill, New York, 1980; Seader and Henley, *Separation Process Principles*, Wiley, New York, 1998; Haas (in Kister 1992)]. Seader (1985) has written an interesting and elegant history of equilibrium-stage simulation equations [Seader, "The B. C. (Before Computers) and A. D. of Equilibrium-Stage Operations," *Chem. Eng. Educ.* **14**(2): 88–103 (1985)].

Real distillation processes, however, nearly always operate away from equilibrium. In recent years it has become possible to simulate distillation and absorption as the mass-transfer rate-based operations that they really are, using what have become known as *nonequilibrium* (NEQ) or *rate-based* models.

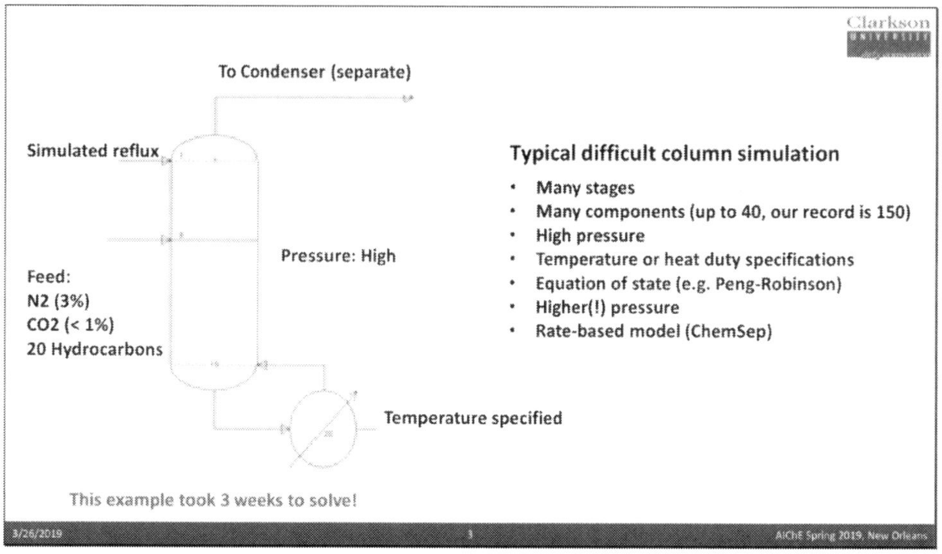

One could make the case that it was the equilibrium model that brought computing into chemical engineering in the first place! One of the incentives for this activity has always been a desire to solve problems with which existing methods have trouble.

This slide shows what, for us, was an especially difficult simulation. This particular simulation has its origin in a column we were modeling for a commercial client. Consequently, nearly all of the details of its operation are confidential. The column configuration shown above is illustrative only. The number of stages shown in the figure is far from the true value.

The aspects of a column model that can lead to a simulation being unusually challenging are summarized at right. All of them were found in the illustrative example that motivates this current work. It took one of us (RT) no less than 3 weeks to solve this problem. That does not mean the computer was running for 3 weeks. It means that it took that long for RT to find an approach that led to a converged solution. In the end he had to use no less than 3 different numerical approaches, one after the other, in order to get a solution. That is, first a highly simplified model of the column was solved first. The results of that simulation were used as a starting point for a second simulation that used a model that was a close approximation to the desired model. The final step was actually several successive cases studies in which the results from the second phase of the solution process were used as starting points for each simulation in the final phase. This involved changing the specification from one that made problem solving easier to the actual specification of interest (the reboiler temperature).

The underlying reasons for the degree of difficulty are:

- The very high pressure leads to difficulties finding appropriate density roots of the PR EOS.
- The liquid flow rate in the bottom of the column was very low. This caused the mass transfer coefficient model and the pressure drop model to give results that we would describe as implausible.

It is simulations like this one – which is far from unusual – that motivated this particular project. In subsequent slides we look at the difficulties posed by these simulations.

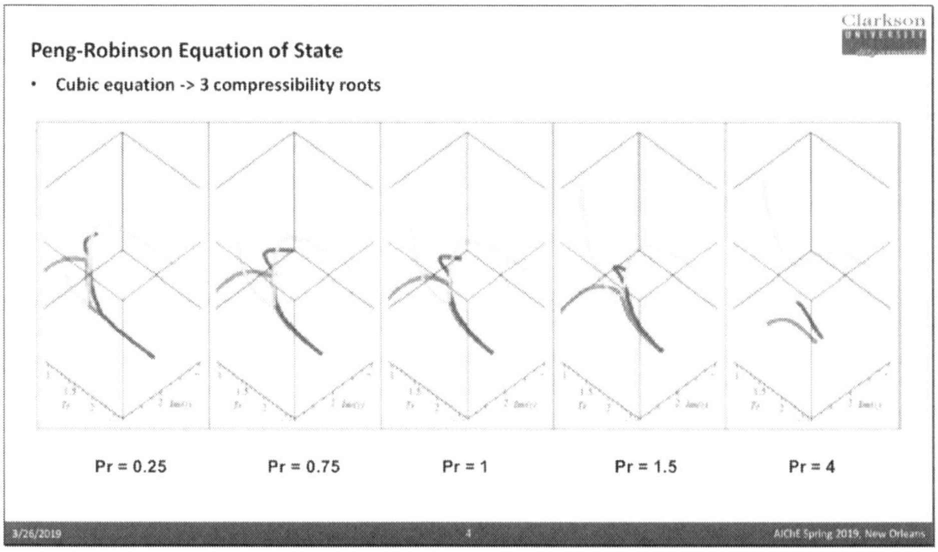

In 1873 Van der Waals proposed a cubic equation of state that serves as an inspiration to thermodynamicists more than 100 years after it was proposed. While the equation has several shortcomings from a theoretical and practical standpoint it remains useful as a starting point for studying other equations of state. An important modification was made by Redlich and Kwong in 1949. The RK EOS was improved by Giorgio Soave and the so-called SRK EOS has become an industry standard. Yet another modification that has become arguably the most widely used cubic EOS was made by Peng & Robinson (PR) in 1976. All of these equations of state – and many more – can be written as a cubic polynomial in terms of density or compressibility. It is important to note that cubic equations always have three roots, but that two of those roots may have non-zero imaginary components under some circumstances. The user of a cubic EOS is unlikely to be able to avoid having to deal with complex roots at some point.

The illustrations on this slide and the next two show the three roots of the PR equation of state as a function of reduced temperature at several reduced pressures. In all cases one of the horizontal axes is reduced temperature. In the 3D plot shown on this slide the other horizontal axis is the imaginary part of the root, the vertical axis is the real part. There must always be one real root, and that is shown by the green line. At low reduced pressures it can be seen that there is a narrow range of reduced temperature where there are three real roots. In this region the largest real root is that of a vapor, the lowest is that of a liquid and the intermediate root has no physical significance. At the lower end of the three real-root region the vapor root bifurcates into the complex domain. At the upper end of this region the liquid root bifurcates into the complex domain. The range of reduced temperature over which three real roots exist decreases in size as the reduced pressure increases. When the reduced pressure is 1 all curves come together at a single critical point where . Above this pressure there appears to be only one real root for all reduced temperatures. At pressures not far above the critical point the real part of the complex roots is greater than the real root over part of the reduced temperature range. However, as the reduced pressure is increased we see that the real root lies above the real part of the complex roots at all reduced temperatures.

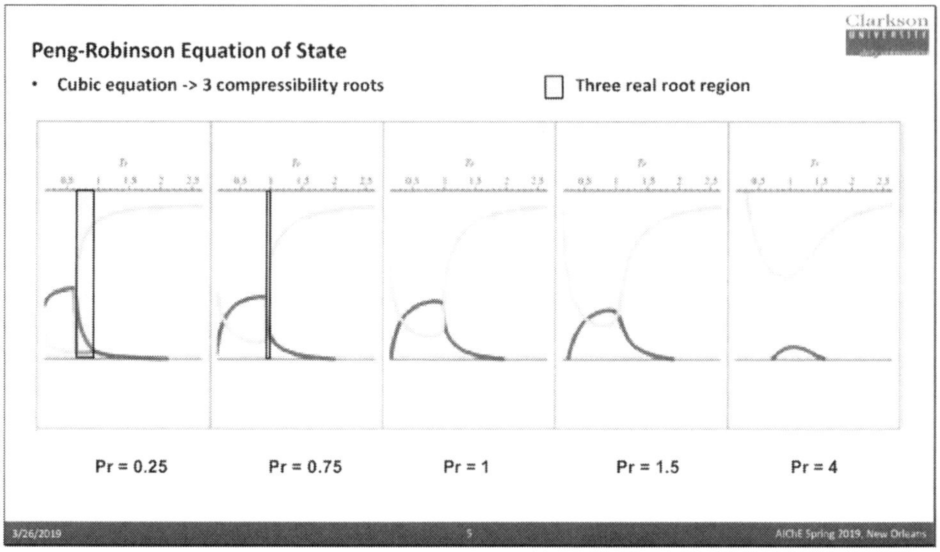

This slide shows a projection of the 3D images on the prior slide so that we are looking through the front left face of the cubes in those images. The two-phase region (where there are 3 real roots) is shown by the pale green shaded area. To the left and right of the two phase region there is only one real root.

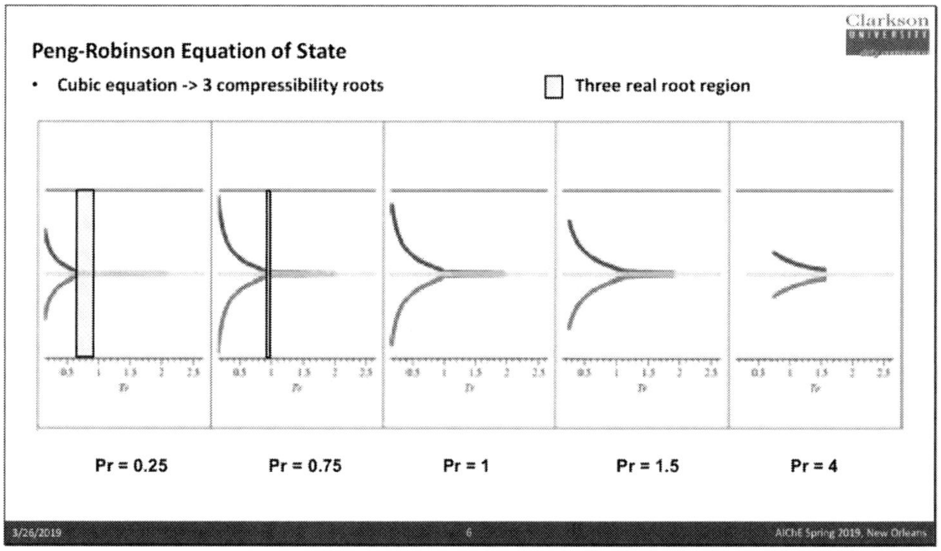

This slide shows a projection of the 3D images on the prior slide so that we are looking through the top of the cubes in those images. The real line is the horizontal line in the center of the diagram. Any curve to the side of the real line has a non-zero imaginary component.

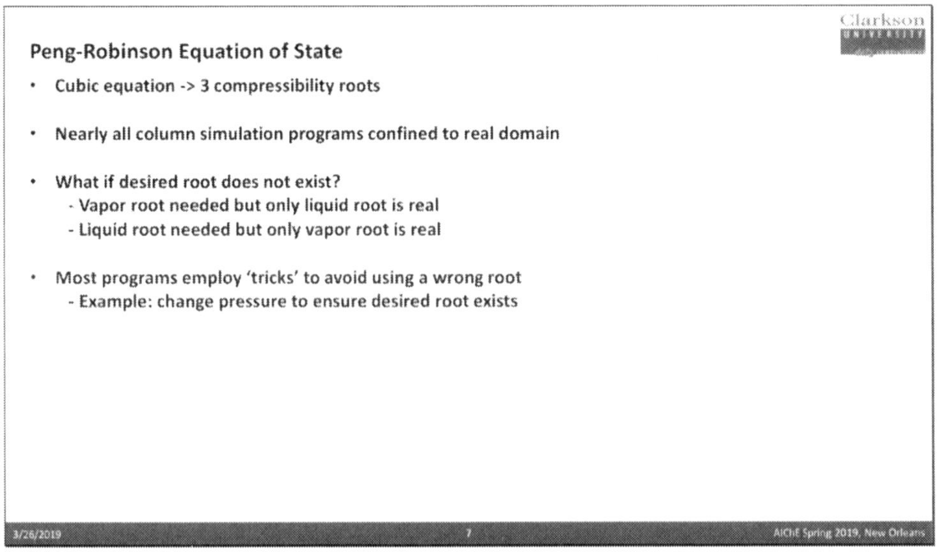

It can be stated with no fear of contradiction that nearly all simulation programs that use cubic equations of state (or other model that has multiple roots) will, from time to time, encounter circumstances in which there is only one real root. It is possible that such occurrences do not cause any difficulty, but what is one to do when a vapor-phase root is desired and only a liquid-like root can be found? Conversely, what if one would like a liquid-phase root but the only one that is real is for a vapor-like phase?

Conventional wisdom is that it is wise to avoid using imaginary roots in process engineering calculations and various strategies have been put forward to avoid these so-called "spurious" roots. Programmers of thermodynamic property packages and flash and distillation algorithms go to great lengths to avoid using a complex root of a cubic EOS. Many of the "tricks of the trade" are proprietary. Poling et al. (Poling, B.E., Grens, E.A., Prausnitz, J.M., 1981. Thermodynamic properties from a cubic equation of state: avoiding trivial roots and spurious derivatives. Industrial & Engineering Chemistry Process Design and Development 20, 127–130. https://doi.org/10.1021/i200012a019) describe a simple method to find and avoid trivial roots and spurious derivatives for a cubic equation of state.

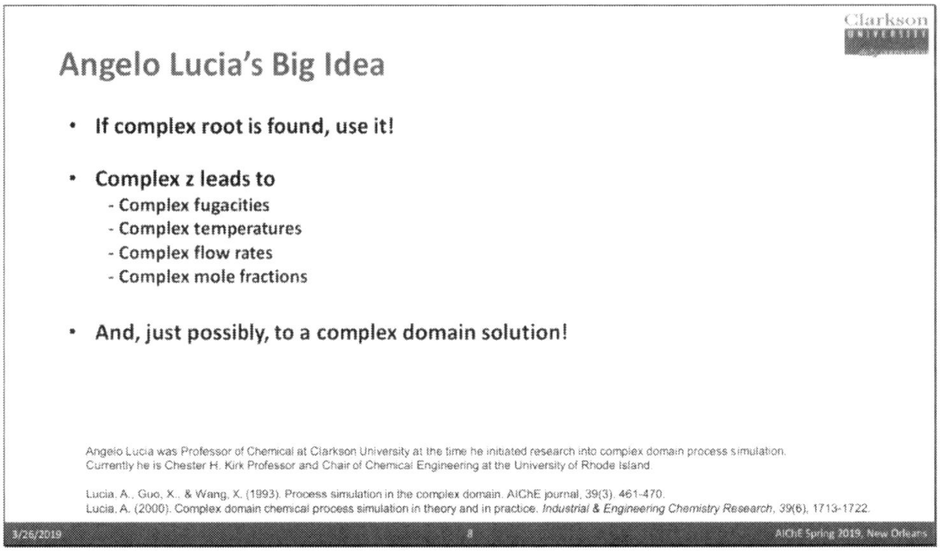

In a series of papers landmark Lucia and coworkers have shown that it can be a good idea to use complex roots during iterative calculations involving cubic equations of state as doing so can help mitigate convergence difficulties that can occur when only the real root(s) are used.

Lucia and Xu (1992) studied the complex domain behavior of various fixed-point methods for solving the Soave-Redlich-Kwong equation of state and show that the bifurcation of a pair of roots into the complex plane coincide with a phase transition. They also show that the dogleg strategy of Powell (1970) can terminate at singular points and that these singular points are saddle points of the complex absolute value function. Based on this observation, they propose an extension of the dogleg strategy for reliably finding real- or complex-valued roots. Lucia and Taylor (1992) have shown that there are some advantages to carrying out equilibrium flash calculations in the complex domain. Although in that exercise, only real valued solutions were obtained, Lucia et al. (1993) obtained complex solutions to dew point temperature calculations when activity coefficient models are part of the phase equilibrium equations. An overview of his work in complex domain process simulation has been given by Lucia (2000).

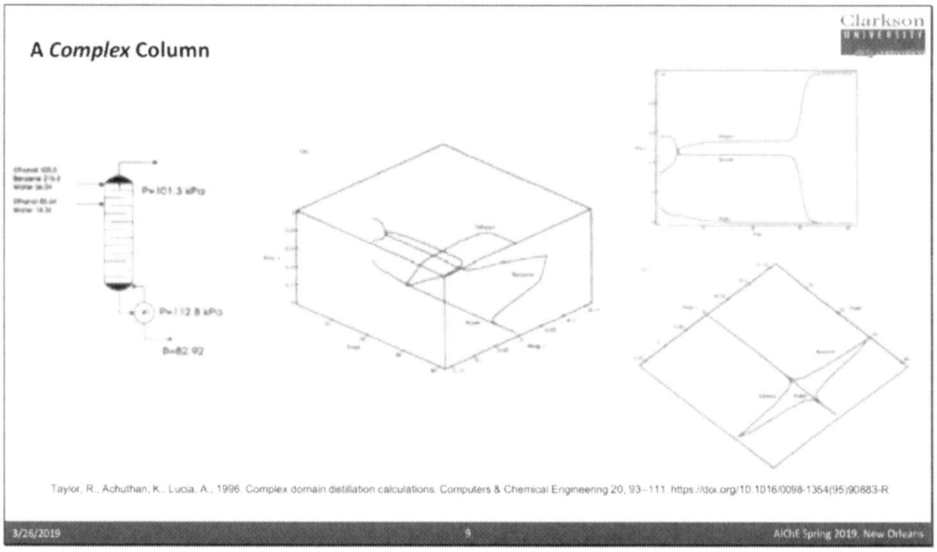

Azeotropic distillation columns can be very difficult to model in the sense that getting the equations to converge can be very hard to do. Even professional programs can have serious difficulties with these problems. Here we discuss the famous problem of separating ethanol (1) from water (3) using benzene (2) as an entrainer.

The column model has 41 stages, a partial reboiler, and **no condenser**. There is a feed containing 85.64 kmol/h ethanol and 14.36 kmol/h water at a pressure of 1.027 bar and a temperature of 311 K to stage 5. In the absence of a condenser we simulate the reflux flow with a second feed to stage 1 containing 103.21 kmol/h of ethanol, 215.8 kmol/h benzene, and 56.29 kmol/h water at a pressure of 1.013 bar and a temperature of 298 K. The bottoms flow rate is 84.0 kmol/h (but see, however, below before you type this value into the proper field). The top and bottom column pressures are 1.01325 bar and 1.128 bar, respectively. The specifications provided here are very similar to a version of this problem studied by Prokopakis and Seider (*AIChEJ*, **29**, 49, 1983).

These calculations are VERY sensitive to the VLE models and parameters. For this simulation the DECHEMA / UNIQUAC q' / Antoine combination of models was used.

It has been shown by several investigators that this particular column model has three different steady state solutions for some sets of specifications. It is this that makes this one of the most famous column simulations.

What is shown in this slide is a solution to this azeotropic distillation column in which some of the mole fractions on some of the stages have imaginary parts that are not zero. This is the first known case in which the solution of a column model resulted in a solution that was partly in the complex domain!

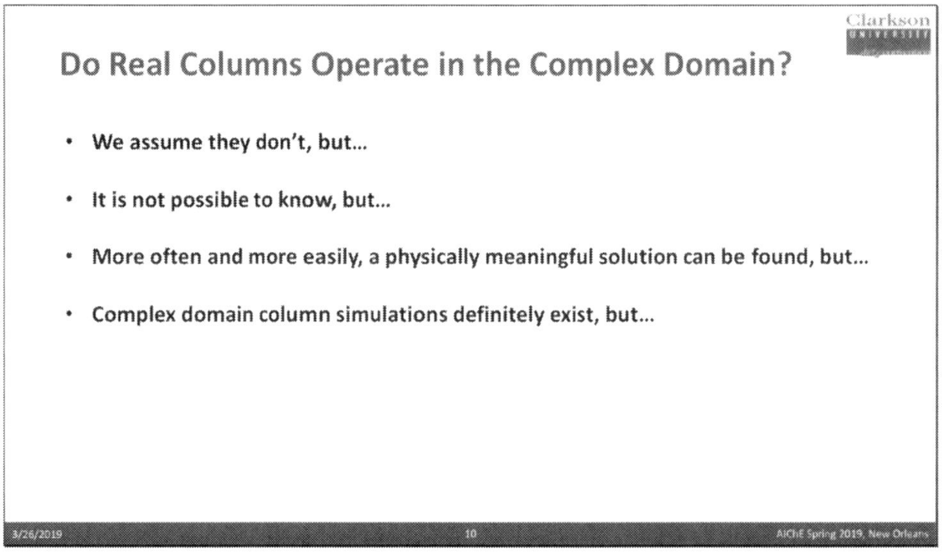

It does not seem plausible that a complex domain solution represents physical reality in anyway. It is, however, impossible to know. Imaginary contributions to mole fractions, temperatures and so on cannot be measured.

Conversely, it is entirely reasonable to assume that a complex domain solution is an artifact of the model. It is a property of mathematics that a cubic equation can have complex solutions. But it is scientists and engineers that choose to use a cubic equation to model the phase behavior of real systems. In doing so we throw out non-physical solutions as meaningless.

We only assign meaning to the imaginary contributions in so far as they help us to find physically real solutions that are entirely within the real domain.

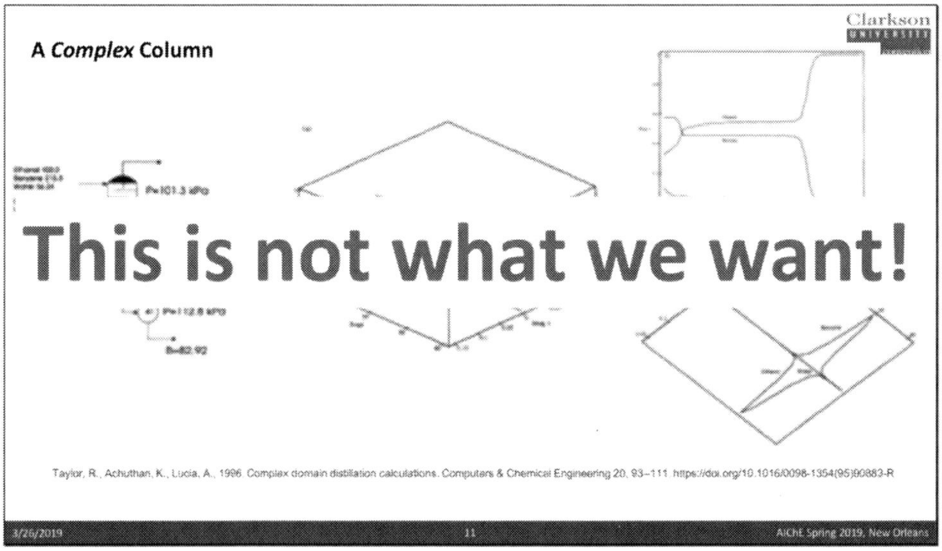

Put another way, complex domain calculations help us to find real domain solutions to simulations that are otherwise exceptionally difficult to converge.

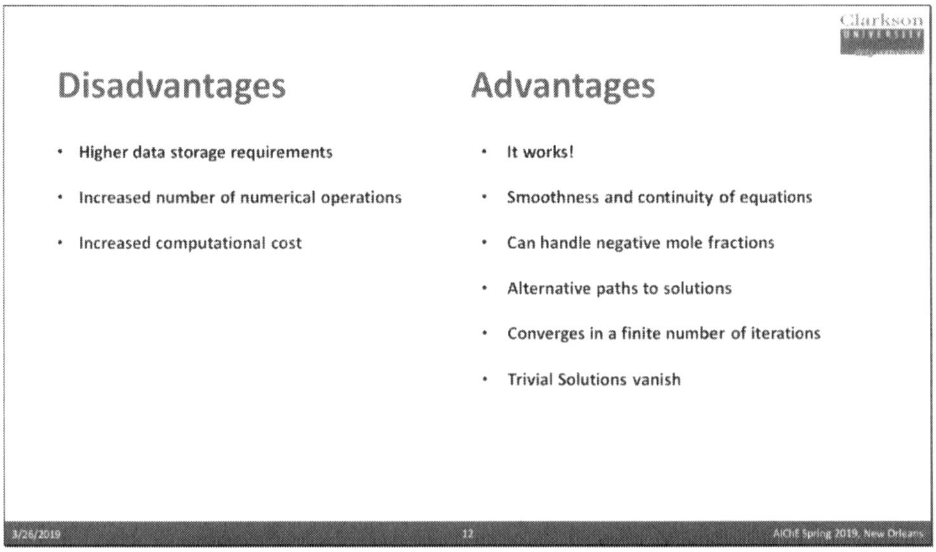

By allowing the iteration path to move through the complex domain, the objective function (conservative equations, physical property models, etc.) remain smooth and continually differentiable, hence creating an uninterrupted path between the initial guess and the solution. The smoothness and continuity is partly due to the fact that complex domain simulation allows for the presence of negative flows, compositions, pressure, temperature, thermophysical properties, etc., that make no sense or are undefined on the negative side of the number line. Thus, removing the need for algorithmic safeguards, which lead to derivative discontinuity.

The complex domain method allows both real and complex valued iterates. Hence, there is a wider area for the algorithm to search for a solution. This makes it more likely to find a real value solution from complex starting points, which cannot be calculated using real valued starting points.

The complex domain method described above guarantees termination of the simulation to a real or complex valued solution, (which provides physically meaningful information), in a finite number of iterations.

Calculation of complex values solution by the algorithm indicates that the solution is infeasible at the desired specification. For example, a complex valued vapour root and a real value liquid root for a TP flash calculation, suggest that the vapour phase does not exist hence, the system is all liquid at the specified temperature and pressure. Lucia and Guo, develop a complex domain based method for determining the feasibility/infeasibility of a process design

It is common for very similar roots of the EOS to be selected for vapour and liquid, especially at high pressure and/or temperature, (where only one physically meaningful root usually exists). This causes subsequent property calculations to be very close to one another and eventual convergence of phase equilibrium calculations to a trivial solution (i.e. $y_i=x_i$). However, when complex valued roots to the EOS are allowed, subsequent property calculations remain different and distinct for liquid and vapour. This in turn leads to the removal of the chance that the algorithm will converge to a trivial solution.

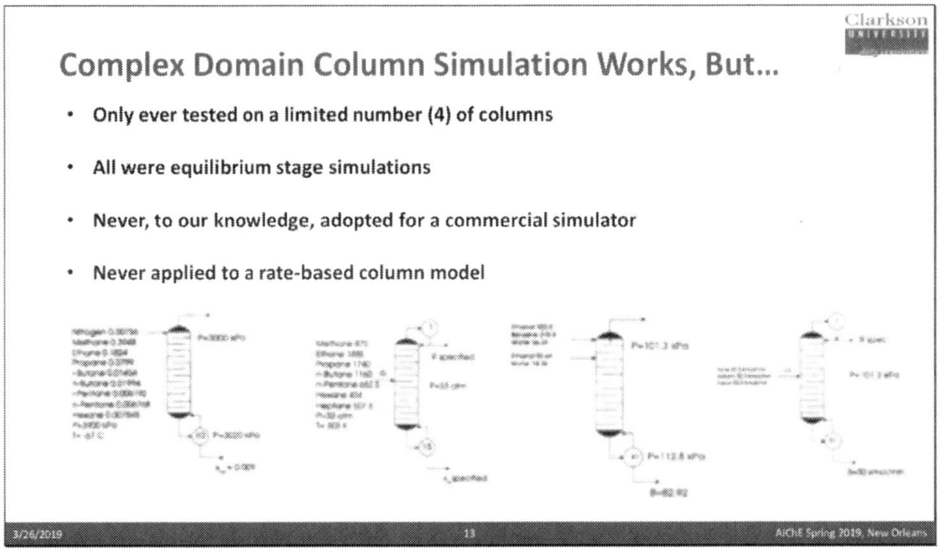

We now know that complex domain simulation are possible and that they have some advantages over real domain calculations, but also a few disadvantages.

While the computational cost of a complex domain calculation is significantly higher than that of a corresponding real domain calculation, there is no valid comparison to be made if the complex domain code actually solves the problem at hand, and the real domain code fails. That is the ultimate reason for using such a method.

Despite the promising methodology put forth by Lucia and coworkers it appears to be the case that the method has never been adopted for use in a commercial simulator. Moreover, the validation of the concept for column simulations was restricted to only four columns, all of which were equilibrium stage models. Also ,as far as we know, the methodology has never been used to solve a rate-based column model.

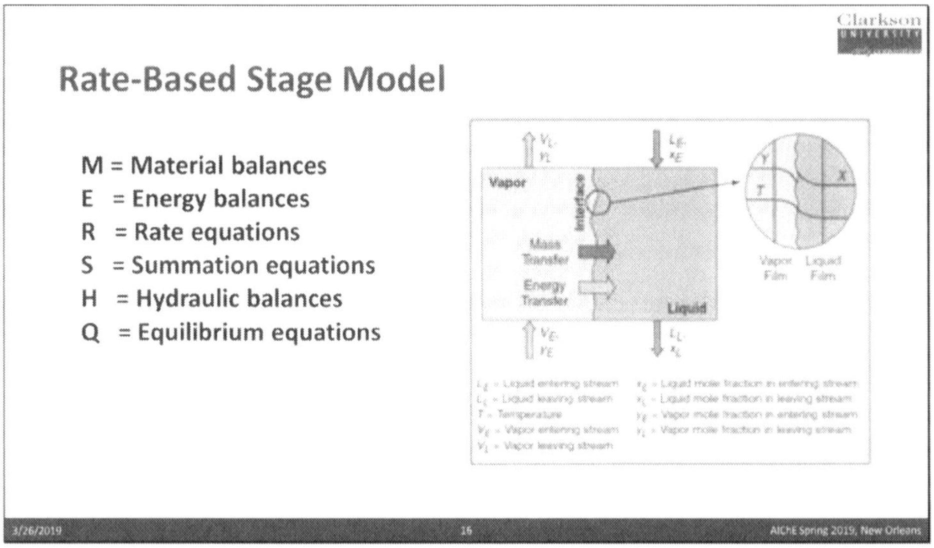

In recent years, a new approach to the modeling of distillation and absorption processes has become available — the so-called nonequilibrium, or rate-based, models. These models treat these classical separation processes as the mass-transfer-rate-governed processes that they really are. The building blocks of the nonequilibrium model shown in the figure are sometimes referred to as the MERSHQ equations, where:

- M represents material balances
- E represents energy balances
- R represents mass- and heat-transfer rate equations
- S represents summation equations
- H represents hydraulic equations for pressure drop
- Q represents equilibrium equations.

Some of these equations are also used in building equilibrium stage models; however, there are crucial differences in the way in which the conservation and equilibrium equations are used in the two types of model. In a nonequilibrium model, separate balance equations are written for each distinct phase. The material balance for each phase includes terms to represent the mass transferred from one phase to the other. For the equation used in the equilibrium stage model, the sum of the phase balances yields the material balance for the stage as a whole. The energy balance is treated in a similar way — it is split into two parts, one for each phase, each part containing a term for the rate of energy transfer across the phase interface.

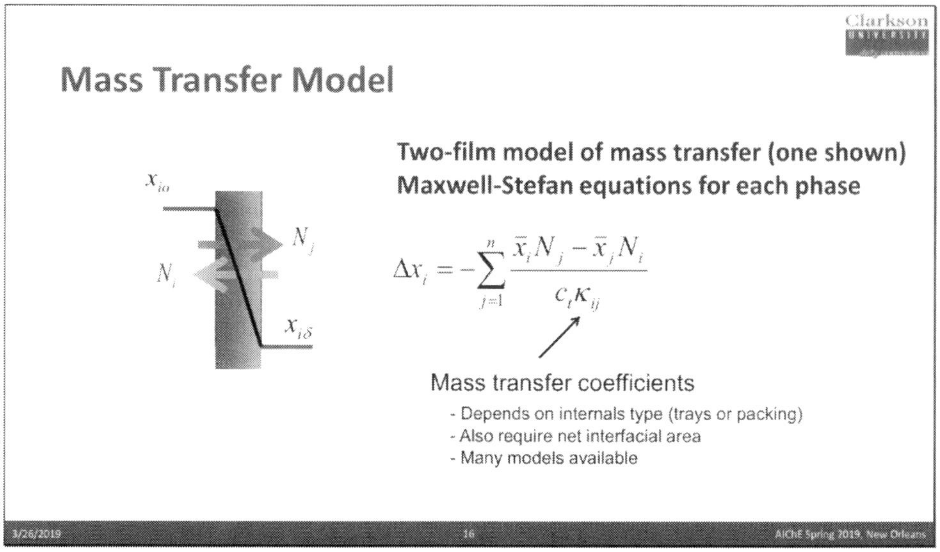

The most fundamentally sound way to model mass transfer in multicomponent systems is to use the Maxwell-Stefan (MS) theory. In our opinion, the MS approach to mass transfer should be what is taught to students, but rarely is that done, even at the graduate level; most texts give little or no serious attention to the matter of mass transfer in systems with more than two components. The MS equations may be derived using nothing more complicated than Newton's second law — the sum of the forces acting on the molecules of a particular species is directly proportional to the rate of change of momentum. The rate of change of momentum between different species is proportional to the concentrations (mole fractions) of the different species and to their relative velocity.

Solving the MS equations can involve the computation of various matrices and functions thereof. In practice, we most often employ a simple film model for mass transfer with a simple difference approximation to the MS equations shown in this slide. Two sets of these equations are needed, one for each phase.

Mass-transfer coefficients and interfacial areas must be computed from empirical correlations or theoretical models. There are many correlations for mass-transfer coefficients in the literature. These coefficients depend on the column design, as well as its method of operation.

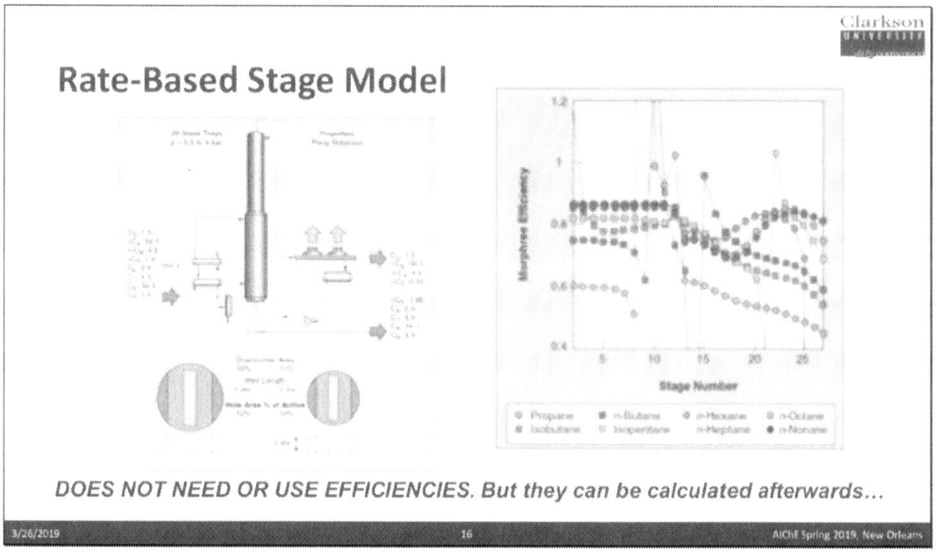

Consider the simple debutanizer shown in this slide. The flowrate and composition profiles do not differ to any significant extent from the results that you would obtain with a conventional equilibrium stage model (although the number of stages and feed stage location would be different). However, a nonequilibrium model can also provide considerable additional information, such as mass-transfer rates and predicted efficiency profiles (the figure on the right). The nonequilibrium model does not use efficiencies; these are computed straightforwardly from the results of a simulation

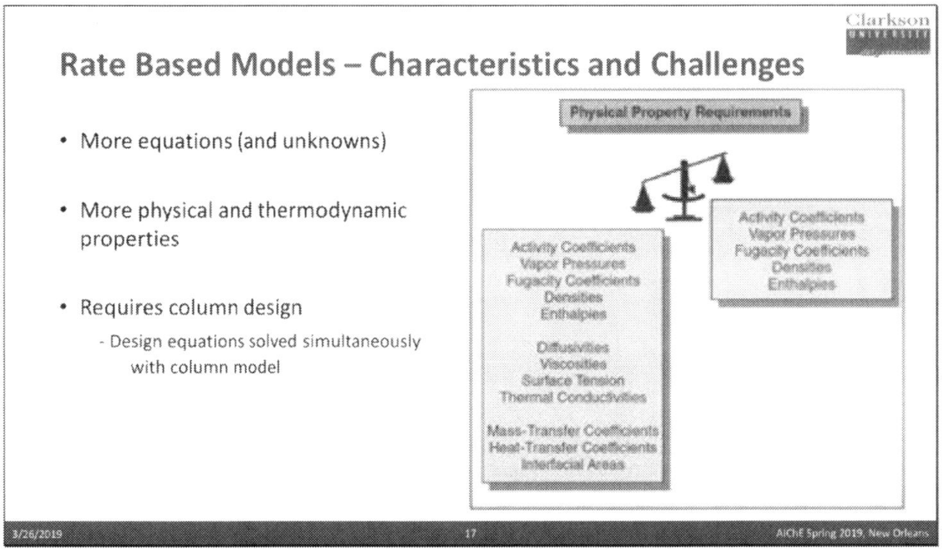

This slide identifies the major physical property requirements. It is obvious that nonequilibrium models are more demanding of physical property data than are equilibrium stage models (except when tray-efficiency or HETP and equipment-design calculations are carried out, but those are done after a simulation and are not needed to carry out the column simulation). The only physical properties required for an equilibrium stage simulation are those needed to calculate the K-values and enthalpies. Those same properties are needed for nonequilibrium models as well.

The estimation of mass-transfer coefficients and interfacial areas from empirical correlations nearly always requires us to know something about the column design. At the very least, we need to know the diameter and type of internal (although usually we need to know more than that, since most empirical correlations for mass-transfer coefficients have some dependency on equipment design parameters, such as the weir height of trays). This need for complete equipment design details suggests that nonequilibrium models cannot be used in preliminary process design (before any actual equipment design has been carried out). However, this is not true. Column design methods are available in the literature, as well as in most process simulation programs. It is straightforward to simultaneously solve equipment sizing calculations and stage-equilibrium calculations. This does not add significantly to the difficulty of the calculations, and it allows nonequilibrium models to be used at all stages of process simulation, including preliminary design, detailed plant design and simulation, troubleshooting and retrofitting. In fact, nonequilibrium models can be particularly valuable in troubleshooting and retrofitting, even to the point of helping identify what particular equipment design detail might be responsible for a column failing to do what it was designed to do.

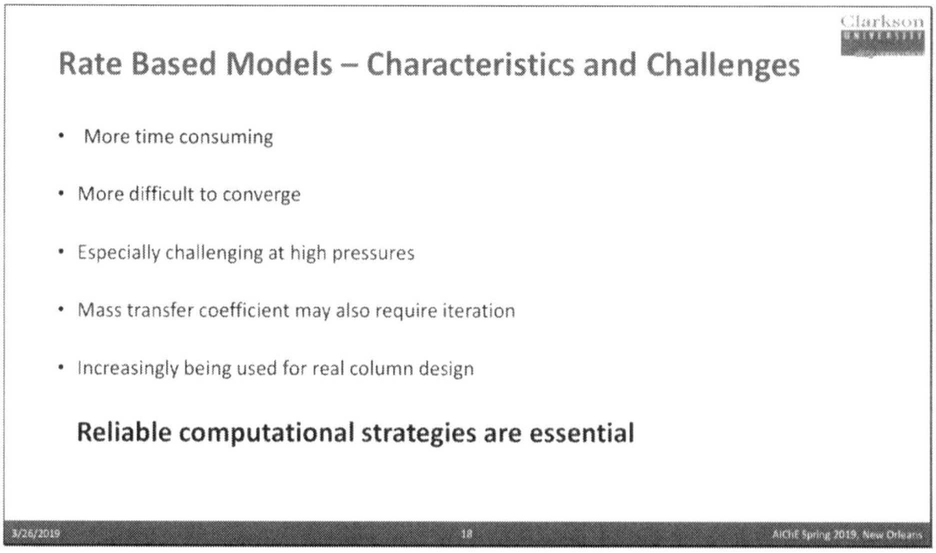

This slide summarizes some of the challenges of rate-based simulation. The fact that this class of model is becoming more widely used for real process design puts an emphasis on developing computational methods that can reliably solve this type of simulation.

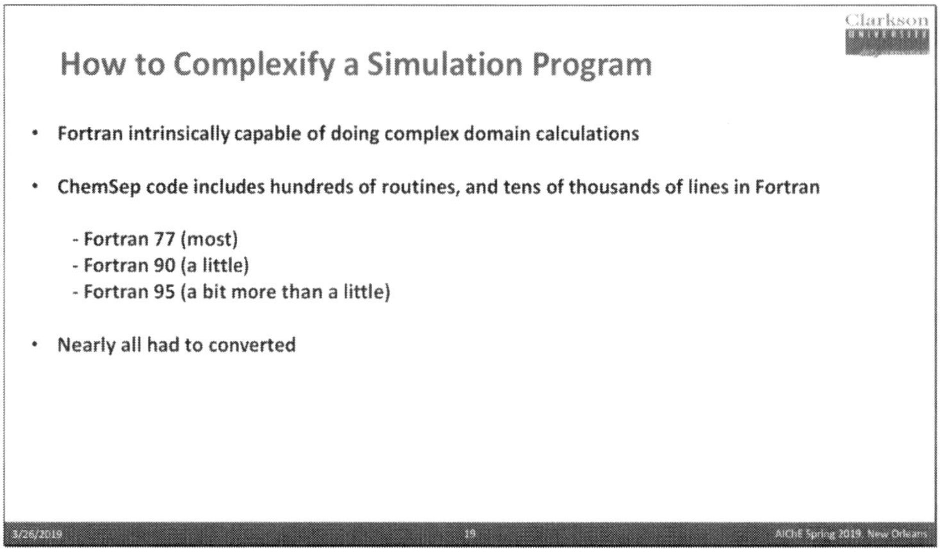

ChemSep is a set of programs that perform multicomponent separation processes simulations. The ChemSep project was started in February 1988 at the University of Technology Delft in the Netherlands, by Arno Haket, Harry Kooijman, and Ross Taylor. ChemSep was designed to be easy to use by students with no experience of engineering software, while having sufficient flexibility and power to appeal to expert users. In pursuit of these objectives ChemSep features a menu-driven, user-friendly interface with an integrated help system and an autopilot mode that leads the novice user through the data input phase. In March of 1991, when the nonequilibrium model was added, the programs and user interface were completely rewritten. Over the years various extensions of the nonequilibrium model were published and added: an extractor model, various dynamic models, a (reactive) cell model, a three phase distillation model, and a total reflux model (for simulating distillation test data). In 2004 a start was made with a new windows-based graphical user interface that we continue to develop. In 2005 ChemSep became Cape-Open compliant. As a result it is now possible to use the program inside flowsheeting tools such as Aspen Plus (Aspentech), PRO/II (SimSci/Escor) and COCO (Amsterchem). In 2006 we made a LITE version of the Cape-Open version of ChemSep that is freely available so from that anybody can test-drive the program \url{http://www.chemsep.com}. ChemSep-LITE is included in COCO \url{http://www.cocosimulator.org}.

The actual calculation program is written in Fortran. We have lost track of the number of lines of code we have written; suffice to say it is a very large number.

There is one aspect of the Fortran language that is of immense value to this project: Fortran has all of the attributes necessary for complex domain calculations. This is not true of all programming languages.

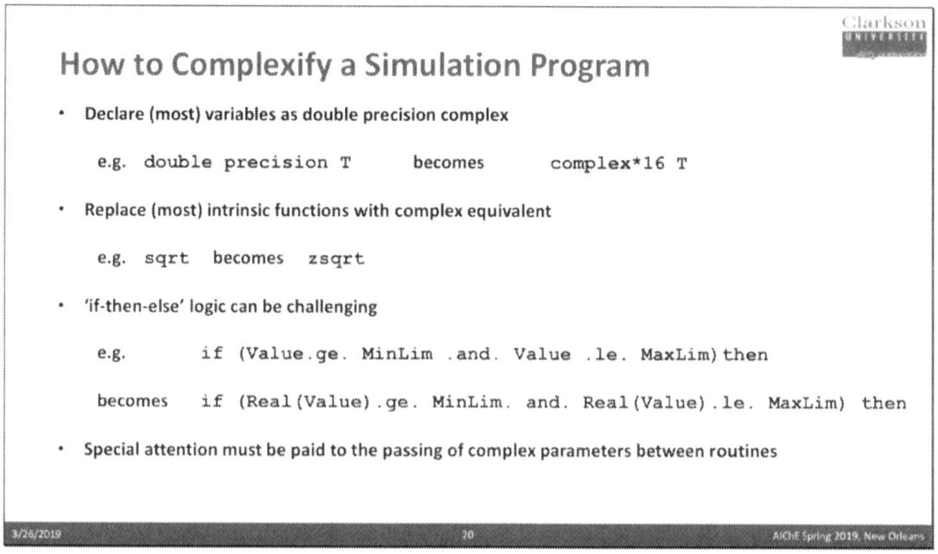

These are the steps we went through to complexify our code. That is, to make it possible to carry out computations in the complex domain. As stated, most variables are converted to complex type. But fundamental property constants (e.g. critical properties, heat capacities. etc.) retain their real values.

Due to the sheer quantity of code, it has taken us a long time to convert all of it and test that each routine does what it is supposed to do. Particularly challenging to convert was the sparse linear equation solver. We use a public domain code called NSPIV. It is always more difficult to modify code written by another party than it is to modify one's own!

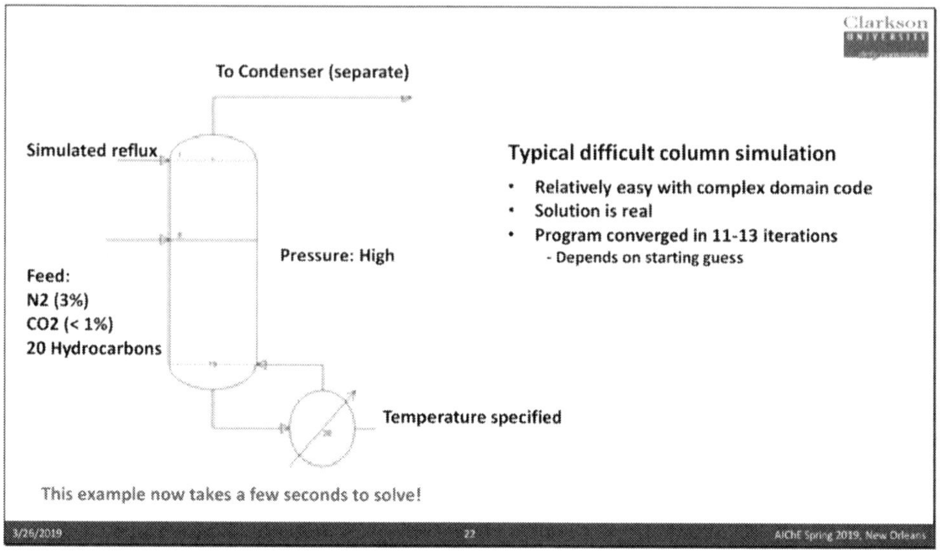

Remember this example from the start of our talk? This is one of the most difficult column simulations we have ever attempted and, as discussed earlier, new strategies were devised simply in order to be able to solve this particular case.

Happily for us, the complex domain method solves this case straightforwardly with no need for multiple steps. Moreover, and very importantly the solution is real. By that we mean that the imaginary part of all variables has become vanishingly small.

Petryschuck Demethanizer-Absorber

- 30 equilibrium stages, with feeds to stages 1, 14, and 22
- Polynomial model for thermodynamic properties
- The column configuration:

	Tray Number	Section Efficiency
Absorbing (top) section	14	87.5%
Center section	9	75%
Stripping (bottom)	7	54%

- Constant pressure of 475 psia.

Petryschuk, W.F., 1964. The Mathematical Simulation of a Three-Feed Demethanizer-Absorber. M.Eng. Thesis, McMaster University.

3/26/2019 23 AIChE Spring 2019, New Orleans

Our complexified code can handle equilibrium stage models as well as rate-based simulations. To illustrate we used our complex domain equilibrium stage model to solve a challenging (for us) simulation of a demethanizer-absorber column from the thesis of Walter Petryschuk (McMaster University, 1964).

Petryschuck Demethanizer-Absorber: ChemSep Model

- 30 equilibrium stages
- Most parameters as Petryschuk, except...
- Peng-Robinson EOS
- Boilup ratio specified

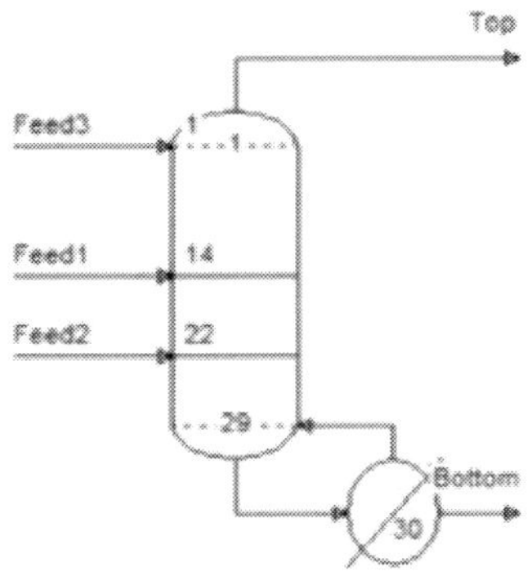

3/26/2019 24 AIChE Spring 2019, New Orleans

We attempted to solve this column with the real domain code in ChemSep. Here is a summary of the column model in ChemSep.

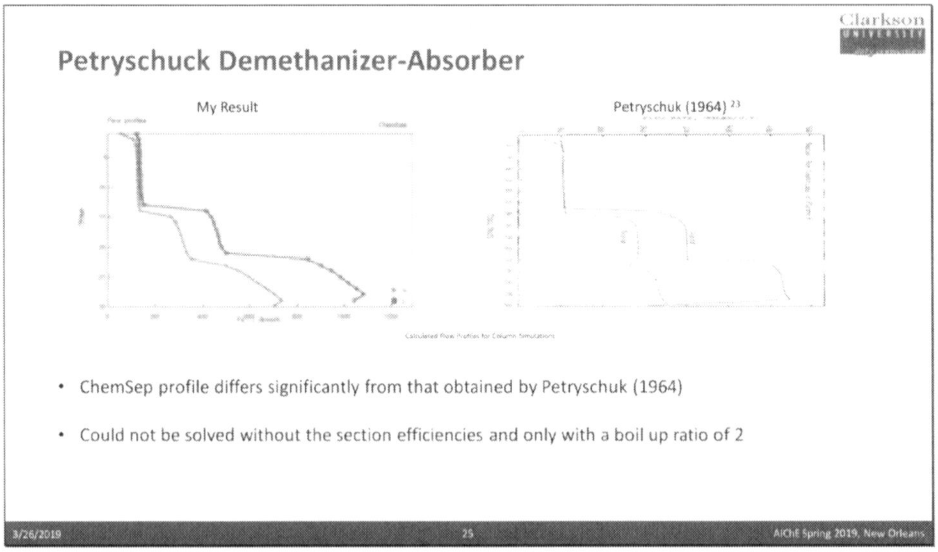

More importantly, the real domain code in ChemSep cannot solve the problem as posed in the thesis unless we specify the section efficiencies and also specify the boil up ratio equal to 2. Any other value and the program failed to find a solution!

The differences between the models are quite considerable. We attribute this mainly to the fundamentally different thermodynamic models.

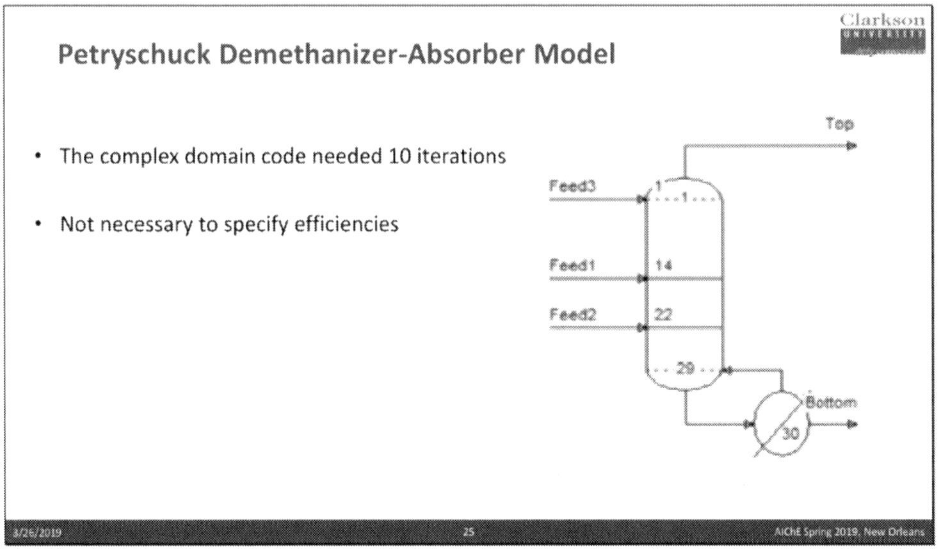

Happily, the complex domain algorithm solves the problem easily for a wide range of parameter values. Efficiencies need to be specified to be less than one.

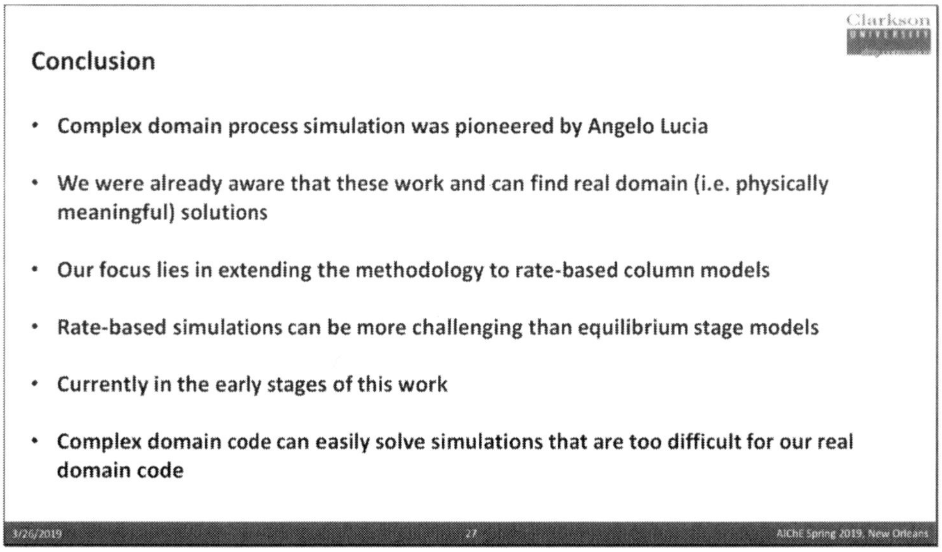

This is as far as we have been able to come. We needed more time than anticipated to complexify the original real domain code so the number of cases we have studied has been limited so far to just 3, two of which we featured in this talk, the third being a large C3 splitter.

What we can say is the complex domain method pioneered by Angelo Lucia works – again.

We still need to carry out a good deal more testing and provide a comprehensive assessment of the value of the methodology. We are, however, confident that complex domain computation will become a valuable additional tool in the arsenal of process simulation experts.

MODELLING OF AN INDUSTRIAL TRICKLE BED REACTOR FOR MAPD SELECTIVE HYDROGENATION WITH CFD ASSISTANCE

Emerentino Quadro, PhD

Process Engineer

Braskem

Brazil

Prepared for Presentation at the 2019 Spring National Meeting

New Orleans, Louisiana, March 31 - April 4, 2019

NOMENCLATURE

Acronyms

CFD	Computational fluid dynamics
FCC	Fluid catalytic cracking
LHHW	Langmuir-Hinshelwood-Hougen-Watson
MAPD	Methyl Acetylene and Propadiene
PR	Peng Robinson thermodynamic model
TBR	Trickle bed reactor
PDID	Pulsed Discharge Ionization Detector
UDF	User defined function

Symbols

A	Pre-exponential factor of equilibrium adsorption constant (m^3/kgmol)
Act	Activity factor
AK	Pre-exponential factor of the reaction rate coefficient (kgmol/m^3s)
Ar	Cross section area (m^2)
C	Adjustable scalar for Marquadt-Levenberg convergence process
D	Catalytic bed diameter (m)
d	Catalyst particle diameter (m)
EA	Heat of adsorption divided by the ideal gas constant (K)
EK	Activation energy divided by the ideal gas constant (K)
F	Force (N)
F_{ij}	Mass flux between phases *i* and *j* (kg/m^3s)
g	Gravitational acceleration vector (m/s^2)
h	Distance from the bottom of the catalytic bed (m)
h1	Length of water column (m)
J	Jacobian matrix
K	Equilibrium adsorption constant (m^3/kgmol)
K_H	Heat conduction coefficient (kg m/s^3K)
k	Reaction rate coefficient (kgmol/m^3s)
L	Length (m)
m_{pq}	Mass flow rate between phases p and q (kg.m^3/s)
p	Pressure (Pa)
Q	Volumetric flowrate (m^3/s)
R_{lg}	Interfacial forces (Pa/m^3)
Re	Reynolds number
r	Rate of reaction (kgmol/m^3s)
S	Vector of parameters increments
T	Temperature (K)
t	Time since the catalytic bed start up (days)
u	Velocity (m/s)
[X]	Concentration of the generic component X (kgmol/m^3)
$\bar{x}$	Average value of variable x
ZA, *ZB* and *ZC*	Adjustable parameters of the deactivation model

Greek letters

ε	Thermodynamic internal energy (m^2/s^2)
λ	Porous medium permeability (m^2)
μ	Viscosity (kg/m s)

ρ	Mass density (kg/m^3)
σ	Standard deviation
$\bar{\bar{\tau}}_p$	Stress tensor in the phase p (Pa)
υ	Mass fraction
Φ	Porosity
Ψ	Normalized standard deviation

Subscripts

H_2	Hydrogen
l	*Liquid phase*
PA	*Propane*
PD	Propadiene
PE	Propyene
MA	Methyl Acetylene
v	Vapor phase

INTRODUCTION

Around 70 million of tons per year of propylene are produced to attend the world demand for polypropylene, acrylonitrile, cumene, acrylic acid, isopropanol, propylene oxide and others oxo-chemicals.[1] This demand is increasing more rapidly than the world's production capacity.[2, 3] Up to 50% of this demand is provided by steam cracking units, while fluid catalytic cracking (FCC) produces 30% of it.[2]

The conventional naphtha steam cracking process produces a C3-cut (stream of hydrocarbons with three atoms of carbon) containing about 85–90 % mol of propylene, close to 5% mol of propane and up to 6% mol of Methylacetylene (MA) and Propadiene (PD).[4] The MAPD is a poison to the catalyst for propylene polymerization and the currently established technology strategy to reach the required MAPD concentration in the propylene polymer grade specification is the selective hydrogenation.[5] The performance of this system is critical to the energy efficiency of the industrial plants and consequently their competitiveness.[6]

At the end of 80's, it was developed the technology to hydrogenate MAPD in Trickle Bed Reactors (TBRs). The main characteristic of the TBRs is the presence of liquid and vapor flowing through a porous medium (fixed catalytic bed), where the vapor is a continuous phase.[7-9] This process offers some advantages, compared to the traditional vapor phase technology, mainly related to reaction selectivity, catalyst life (due to the removal of produced oligomers from the catalyst surface by the liquid flow) and equipment costs, once the required reactor volume is smaller than to vapor phase.[10-12] Despite of the increasing using of MAPD hydrogenation in liquid phase, the studies regarding this reaction, considering both the kinetic aspects and flow distribution, are quite rare in the literature.[4] The kinetic models for the vapor phase MAPD hydrogenation reactions have been studied for various authors; while the kinetic modeling of liquid phase hydrogenation of MAPD is seldom seen in academic papers or public reports.[4]

Uygur et al. studied the MAPD hydrogenation kinetics in liquid phase in 1998.[13] Reaction rates were based on a mechanism derived from the LHHW model. However, these results were obtained from laboratory tests and scale up of TBR usually results in a poor design. In addition, this model was developed for a narrow range of variation of the handled variables.

Other study about kinetics of MAPD hydrogenation in liquid phase was published by Yu et al in 2002.[14] They used simple models, based on the Arrhenius equation, and various simplifying assumptions related to the dynamics of the flow in the porous medium.

In 2011, Wu et al.[4] implemented an interesting and complete study to determine the hydrogenation kinetics of MA and PD separately in liquid phase, by using LHHW-based models adjusted to industrial plant data. However, many simplifications were assumed: The reactor simulation model was a plug flow type, it was not considered the effects of radial or axial mass dispersion, no interactions between phases, the solid phase (catalyst) was ignored and it was assumed there was no pressure gradient.

In 2014, Samimi et al.[8] investigated the application of an empirical statistical model to optimize an industrial TBR for MAPD hydrogenation. This study was based on response surface methodology. The design of experiments obeyed a multilevel factorial, performing a total of 192 experiments. Once the model was non-phenomenological, its applicability was limited and it was not possible to analyze it from the point of view of reaction mechanism.

A relevant difficulty to be considered in order to identify the intrinsic kinetic reaction rate models for TBRs is the radial variation of velocity, temperature and composition due the mass, momentum and energy transport limitations.[10, 15, 16] Therefore, a realistic model for this reactor should take these effects into account. CFD (Computational Fluid Dynamics) has been applied successfully to represent these effects inside of many industrial equipment,[17, 18] including TBRs,[19-22] for being a set of numerical resources used to describe the momentum, energy and mass transfer, associated with fluid flow and its interactions with the geometries.[23, 24]

It is therefore not possible to find in the literature a general and realistic phenomenological intrinsic kinetic model to predict the behavior of a industrial-scale MAPD hydrogenation reactor in liquid phase. It is the objective of this contribution to fill this lacuna. The kinetic models parameters regression was performed in an iterative algorithm by using CFD coupled to mass transfer between phases and kinetic models, adjusted to fit the industrial operational data. A separate set of experimental data were used to validate the models. Based on the obtained model, it was possible to find the optimal operation conditions to maximize the catalytic bed selectivity and run length. The findings also allow us to understand the phenomena inside the TBRs applied to MAPD hydrogenation and to design new TBRs to this function.

METHODS

Process Description

The running on-site MAPD converter process in liquid phase is described schematically in the Fig 1. In this figure, the instruments used to monitor the temperature are indicated as **T**, **F** represents the flow rate meters, the instruments used to indicate the pressure are signalized with **P** and **A** indicates on line analyzers. The operator is free to set the values of recirculation flow rate (*F3*) and the amount of hydrogen fed to the reactor (*F2*). Both reactors are equipped with a fixed catalyst bed, through which flows a double phase stream, typical of the TBR. Palladium (Pd) dispersed on high purity alumina is the active metal present in the most of the catalyst for selective alkynes and alkenes hydrogenation, including MAPD, to the corresponding olefins.[25] The focus of the present investigation is the first MAPD hydrogenation reactor.

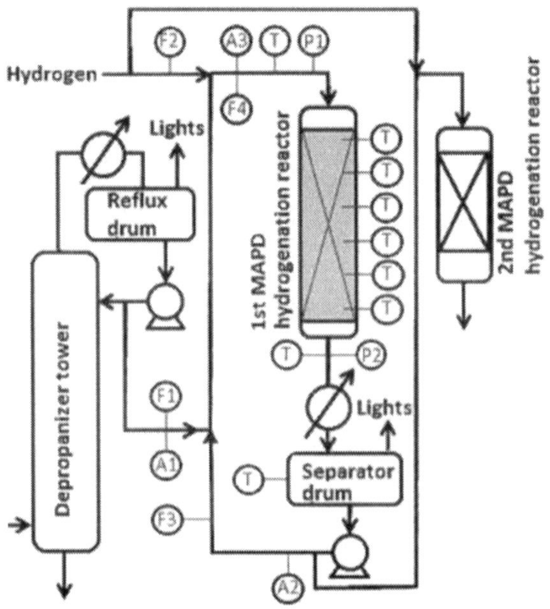

Figure 1. Schematic drawing of the MAPD hydrogenation system and its instrumentation.

Methodology summary

Figure 2 shows a diagram with the main steps of the applied methodology. The number of cases (a set of operational data representing the industrial system in a certain moment along of the catalytic bed run length) utilized to the kinetic models parameters regression were 38. Other 10 cases were utilized for the model validation. In the next topics, the main steps of this methodology are described in detail. The startup points have been highlighted in black box.

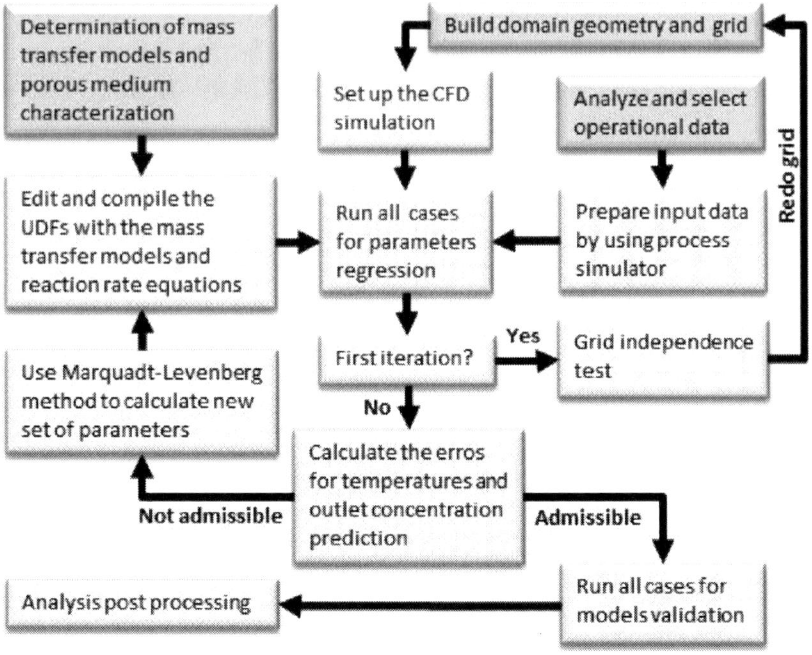

Figure 2. Simplified diagram of the methodology.

Operational data analysis

Thirty-eight cases of steady state operation condition were selected, whose operational data could be used in the kinetic model regression procedure. Definition of steady state is when operating conditions are not changing, for example, variations of feed flow, inlet or outlet MAPD concentration, hydrogen feed flow etc. The methodology consisted of collecting, at least, four consecutive values of the variables measured every 30 minutes. The standard deviation (σ) should be equal or lower than 2% of their average value ($\bar{x}$), as shown in Eq. (1), where ψ is the normalized standard deviation.

$$\psi = \frac{\sigma}{\bar{x}} = \frac{\sqrt{\frac{\sum_{i=1}^{k}(x_i-\bar{x})^2}{n-1}}}{\bar{x}} \leq 0.02 \qquad (1)$$

Another criterion to choosing the set of operational data to be used in the simulations was the MA and PD concentrations in the reactor outlet, they should be higher than 100 ppm in order to viabilize the solution of the equations system. The Madon–Boudart principle[26] says that, in the absence of all transport limitations, the reaction rate is proportional to the number of active sites. At same process conditions (inlet temperature, inlet concentration of reactive species, feed flow and pressure), we have observed a consistent increasing of outlet reactive species concentration (MA, PD and H2) during the catalyst bed campaign, the pressure drop is within the predicted value and it does not vary significantly along the time (it is conductive to believe that eventual flow maldistribution is not big enough to affect the overall reaction rate). Besides, laboratory analysis performed by the catalyst manufacturer on the exhausted catalyst (by using a lab scale reactor at very low residence time –a differential reactor) has showed a lower catalytic activity than the virgin catalyst. Based on these points, we concluded the overall reaction rate is mainly governed by the catalyst activity (besides, of course, temperature and reactive species concentration). Even though all these information, it is not possible to guarantee the absence of transport limitation. For this reason, the activity factor (Act, presented in the next section) will take account, besides the catalyst deactivation, any transport effect responsible to affect the overall reaction rate. Based on this, it was possible to isolate the intrinsic kinetic of the MAPD hydrogenation reaction[21].

Reactions and kinetic models

The main reactions that take place in this reactor are the MAPD hydrogenation, Eqs. (2) and (3), besides of the undesired propylene hydrogenation, Eq. (4), where their corresponding enthalpy variations[4] are represented.

$$MA + H_2 \xrightarrow{r_{MA}} Propylene \qquad \Delta H = -164.9\ kJ/mol \qquad (2)$$

$$PD + H_2 \xrightarrow{r_{PD}} Propylene \qquad \Delta H = -171.6\ kJ/mol \qquad (3)$$

$$Propylene + H_2 \xrightarrow{r_{PE}} Propane \qquad \Delta H = -124.1\ kJ/mol \qquad (4)$$

The most accepted mechanism for these reactions, for both liquid and vapor phases, is based on the concepts of Langmuir-Hinshelwood-Hougen-Watson (LHHW), which is applicable for heterogeneously catalyzed reactions when both reagents molecules are adsorbed in catalytic sites [6]. This mechanism results in the equations Eqs. (5), (6) and (7), without the deactivation correction, where [*X*] represents the concentration of the generic component *X*.[4]

$$r_{MA} = \frac{k_{MA}K_{MA}K_{H_2}[MA][H_2]}{(1+\sqrt{K_{H_2}[H_2]}+K_{MA}[MA]+K_{PD}[PD]+K_{PE}[PE]+K_{PA}[PA])^3} \quad (5)$$

$$r_{PD} = \frac{k_{PD}K_{PD}K_{H_2}[PD][H_2]}{(1+\sqrt{K_{H_2}[H_2]}+K_{MA}[MA]+K_{PD}[PD]+K_{PE}[PE]+K_{PA}[PA])^3} \quad (6)$$

$$r_{PE} = \frac{k_{PE}K_{PE}K_{H_2}[PE][H_2]}{(1+\sqrt{K_{H_2}[H_2]}+K_{MA}[MA]+K_{PD}[PD]+K_{PE}[PE]+K_{PA}[PA])^3} \quad (7)$$

The reaction rate coefficients are based on the Arrhenius equation and they are described by the Eqs. (8), (9) and (10). A temperature parameterization was carried out in order to facilitate the parameter estimation, where 310 K is the average reactor temperature profile.

$$k_{MA} = AK_{MA}\exp(-EK_{MA}\left(\frac{1}{T}-\frac{1}{310}\right)) \quad (8)$$

$$k_{PD} = AK_{PD}\exp(-EK_{PD}\left(\frac{1}{T}-\frac{1}{310}\right)) \quad (9)$$

$$k_{PE} = AK_{PE}\exp(-EK_{PE}\left(\frac{1}{T}-\frac{1}{310}\right)) \quad (10)$$

The equilibrium adsorption constants for each component are described by the Eqs. (11)-(15), where a temperature parametrization was also carried out.

$$K_{H_2} = A_{H_2}\exp(EA_{H_2}\left(\frac{1}{T}-\frac{1}{310}\right)) \quad (11)$$

$$K_{MA} = A_{MA}\exp(EA_{MA}\left(\frac{1}{T}-\frac{1}{310}\right)) \quad (12)$$

$$K_{PD} = A_{PD}\exp(EA_{PD}\left(\frac{1}{T}-\frac{1}{310}\right)) \quad (13)$$

$$K_{PE} = A_{PE}\exp(EA_{PE}\left(\frac{1}{T}-\frac{1}{310}\right)) \quad (14)$$

$$K_{PA} = A_{PA}\exp(EA_{PA}\left(\frac{1}{T}-\frac{1}{310}\right)) \quad (15)$$

In normal operation conditions, MA and PD react with themselves resulting in oligomers, called green oil [4]. Part of these oligomers do not leave the reactor; they accumulate over the catalyst sites and work as a deactivating agent. In typical industrial process conditions, the rate of green oil formation is much smaller than the other reactions and it is function of the temperature and MAPD concentration, therefore the catalyst deactivation rate is function of these factors. This work proposes the non-phenomenological deactivation rate equation shown in Eq. (16), where *ZA*, *ZB* and *ZC* are adjustable parameters. This empirical model assumes that the activity factor (*Act*) does not vary radially. Besides, it correlates the activity with the distance from the bottom of the catalytic bed and the time of operation since the start up.

$$Act = 1/(1 + \frac{\exp(ZA.h.t^{ZB})}{ZC}) \quad (16)$$

The proposed kinetic models coupled with a catalyst deactivation model were fed to the CFD simulation by UDF (User Defined Function), a compiled C code utilized by the CFD solver to compute the effect of the reactions on the mass balance per component and the heat generation in each cell of the grid.

Porous medium characterization

When a liquid reactive mixture is fed to a catalytic bed, it is mandatory to consider the effects of the porous medium, in order to account properly the mass, momentum and energy transport effects.[27] It has been adopted in this work, due to the computational overhead associated with CFD simulations, a pseudo homogenous approach to represent the porous medium. This model ignores the liquid flow at the particle scale and assumes all particles are completely wetted. It has been shown that this approach is satisfactory to predict trickle bed performance and its dependency on the process variables.[28]

For CFD simulation, the porous medium model adds a momentum sink in the governing momentum equations, and it reduces the available area to the fluid flowing.[29] The turbulence models for porous medium are only approximated; however, once the expected turbulence inside the catalytic bed is not high, this is not a serious restriction for the present work.

The main parameters employed in porous medium characterization for the CFD steady state simulation are the permeability and the void fraction[20]. In this context, this research has determined experimentally the permeability (λ) of a catalytic bed currently used in a trickle bed reactor for selective hydrogenation of MAPD by using the apparatus schematically shown in Fig 3. The porosity and the bulk density of the catalytic bed were also measured.

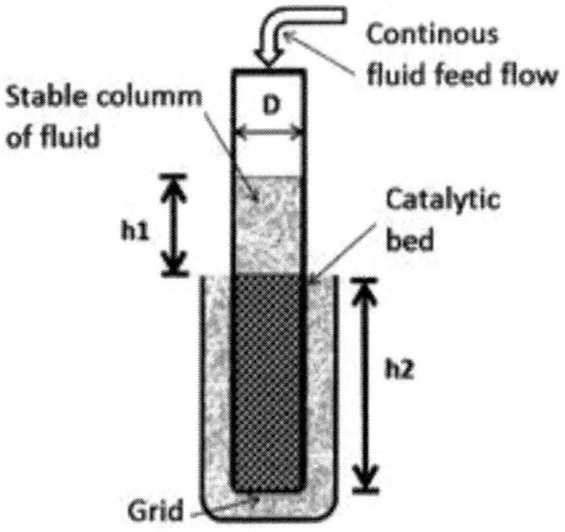

Figure 3. Diagram of the apparatus utilized to measure the catalytic bed permeability.

The laboratory method used demineralized water flowing through a catalytic bed loaded in a glass tube partially immersed in a recipient with water. The main experimental parameters were the water flowrate (*Q*), the hydraulic gradient (pressure drop, related to *h1*) and the geometric characteristics of the bed (cross section area, *Ar*, and the bed length, *L*). The values of these parameters are shown in Table 1. In order to avoid wall effects, the ratio *D/d* must be higher than 15,[27, 30] where *D* is the glass tube diameter and *d* is the catalyst particle diameter (3.13 mm, in a Gaussian distribution).

Table 1. Main parameters of the experiments.

Parameter	Value
Water temperature (°C)	26.0
Water mass density (kg/m^3)	999.60
Water viscosity (kg·m^{-1}·s^{-1})	0.00095
Bed cross section area (m^2)	0.00694
Bed height (m)	0.228
Bed diameter (m)	0.094

The relation between these variables, described by the Eq. (17), was established by Henry Darcy. This equation is strictly valid for Reynolds number lower than 10 and single phase flowing through a porous medium.[20, 31, 32] Despite this, the obtained permeability was used in the CFD simulation, by adjustment of an empirical model, correlating the permeability and superficial velocity. This way, it was possible to estimate the pressure drop in the TBR.

$$Q = \frac{\lambda\, Ar}{\mu} \frac{\Delta p}{L} \qquad (17)$$

Mass transfer models

The prediction of the hydrogen concentration in reactive liquid mixture at various temperatures is an important factor to the representativeness of the CFD simulation for hydrogenation of MAPD in the TRB. The model of thermodynamic equilibrium proposed by H. G. Grayson and G. W. Streed in 1963 has been widely used for determining the solubility of hydrogen in systems containing non-polar hydrocarbons with temperatures below 250°C (482°F) and pressure lower than 9900 kPa (1435.9 PSI). However, depending on the average molecular weight of hydrocarbons present in the liquid phase, even this model has been shown deviation of up to 5% [33]. Therefore, due to the difficulty of thermodynamic models to represent the concentration of hydrogen in liquid phase of light hydrocarbons, experimental tests were carried out to determine an empirical for the hydrogen absorption, at equilibrium, by a mixture of hydrocarbons with three carbon atoms (propane, propylene, MA and PD).

The experimental apparatus utilized to evaluate the hydrogen C3-cut vapor-liquid equilibrium is shown schematically in Fig 4. This system (vessel and piping) are thermally insulated. These tests were performed at 2015 kPa (292.2 PSI), operating pressure of the on-site industrial TBR, and the temperatures varied from 26 to 46°C (78.8 to 114.8°F). The hydrogen concentration in the liquid phase for each temperature was measured by Gas Chromatography in triplicate using a one-valve analyzer with 1/8-inch packed column, equipped with a ten-port valve for back-flushing. The columns installed in this analyzer were 6 ft Hayesep Q 80/100 mesh and 6 ft Shincarbon Q 80/100 mesh with a PDID (Pulsed Discharge Ionization Detector) detector.

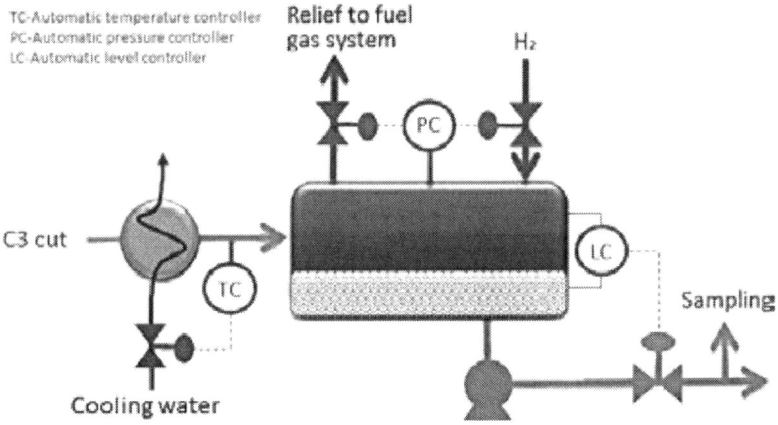

Figure 4. Schematic representation of the equipment utilized to evaluate the hydrogen C3-cut vapor-liquid equilibrium.

To make sure that the system was in steady state, the experimental data were considered valid only in case the absolute pressure and the absolute temperature values of the system had shown less than 0.1% variation during, at least, 2 hours. Besides, the maximum variation of the obtained values for mass fraction of hydrogen in liquid phase should be less than 1.0%

The obtained empirical model was fed to the CFD simulation by using a UDF to compute the hydrogen mass transfer rate between phases in each cell of the grid.[28]

The mass transfer due to the C3-cut vaporization was taken into account by using an empirical model developed from a thermodynamic simulation (performed by using a commercial software), based on the Peng-Robinson model (Equation of State), which is widely used in the hydrocarbon industry and accurately represents the relationships between temperature, pressure and phase compositions in binary and multicomponent systems.[34] Again, this model was fed to the CFD simulation by using an UDF.

Domain geometry

The main dimensions and parts of the on-site running TBR used to the MAPD selective hydrogenation are represented in the Fig 5.

The domain geometry consists in three different parts. The body 1 corresponds to the space between the base of the feed distributor and the top of the catalytic bed; the body 2 refers to the space occupied by the catalytic bed; and the body 3 represents the space between the base of the catalytic bed and the outlet nozzle. These divisions were required in order to attribute the properties of porous medium and reaction rate to the catalytic bed (body 2) in the CFD simulation. It was assumed the reactor feed flow is equally distributed among all chimneys of the feed distributor (the distributor was tested in pipe shop by using water and the flow rate from each chimney was measured, as result, it was confirmed the uniformity of the flow through each chimney), therefore the fluid flow in the top of the body 1 is not perfectly uniform, the CFD simulation has assumed punctual flow from the places where each chimney is localized.

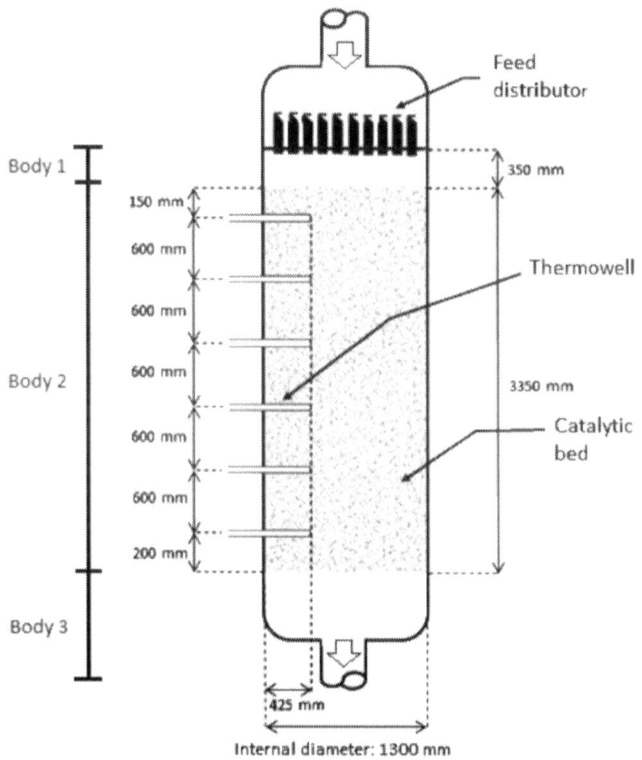

Figure 5. Main parts and dimensions of the reactor.

CFD MODELLING AND NUMERICAL APPROACH

The CFD calculations were performed in the Finite Volume approach by using a commercial software in steady state mode, 3D, multi-phase (Volume Of Fluid – VOF model - was applied) and double precision. The VOF model is a surface-tracking approach considering a fixed Eulerian grid, where a set of momentum equations is shared by both fluids, besides the volume fraction of each of the fluids is tracked for each cell of the domain.[34, 35] A pressure-based Navier-Stokes solution algorithm was employed. The turbulence model applied was the k-ε, largely used for porous media,[20, 36] with realizable parameters, listed in Table 2.[29]

Table 2. Parameters utilized in the turbulence model (k-ε).

Parameter	Value
$C_{1\varepsilon}$:	1.44
C_2:	1.90
σ_k:	1.00
σ_ε:	1.20

A non-structured 3D grid, recommended for complex geometry,[23, 37] was built by using a commercial software. The criteria adopted to ensure minimum grid quality are listed in Table 3.

A grid sensitivity analysis was performed and confirmed that a higher resolution discretization would not affect the results. To confirm the grid-independence from velocities profiles under specified operation conditions, it was verified that the simulations results varied less than 1%, after the number of cells was increased by 50%.

Table 3. Grid quality criteria.

Parameter	Reference
Skewness:	<0.60
Aspect ratio of tetrahedral cells:	>0.40
Minimum angle:	>18°
Orthogonal quality:	>0.40

The Navier-Stokes equations, Eqs. (18)-(20), were solved simultaneously in an iterative process to achieve a final solution. However, other equations (reaction rates, mass transfer, turbulence etc.) were associated until a suitable description of the physical process to be produced. The maximum scaled residual (calculated by Fluent) accepted in simulation for mass imbalance (continuity), momentum and turbulence parameters (k and ε) were 10^{-4}. However, this was not enough to guarantee that the convergence has been reached. Once achieved this residual, other 10^3 iterations were run to ensure the variation of the velocity in a same point was lower than 1%.

$$\frac{\partial \rho}{\partial t} + \nabla.(\rho\, u) = 0 \tag{18}$$

$$\frac{\partial u}{\partial t} + (u.\nabla)u = -\frac{1}{\rho}\nabla p + F + \frac{\mu}{\rho}\nabla^2 u \tag{19}$$

$$\rho\left(\frac{\partial \varepsilon}{\partial t} + u.\nabla\varepsilon\right) - \nabla.(K_H \nabla T) + p\nabla.u = 0 \tag{20}$$

The main boundary conditions are listed below:

- The total feed flow, liquid and vapor, enters in the domain equally through all chimneys of the feed distributor;
- Both phases fed to the reactor are in vapor-liquid equilibrium;
- The reactor was considered perfectly adiabatic;
- The catalytic bed was treated as an isotropic porous media;
- The simulation was run in steady state mode.
- The no slip condition was assumed;
- The total outflow is located on the bottom of the reactor.

For this work, the mass conservation equation for each phase is represented by Eq. (21).

$$\frac{\partial(\phi\rho_p\alpha_p)}{\partial t} + \nabla.\left(\phi\, \rho_p\alpha_p u_p\right) = m_{qp} - m_{pq} \tag{21}$$

where ρ_p, α_p and u_p represent the mass density, volume fraction and mean velocity, respectively, of phase *p* (l or v), ϕ is the porosity of the porous media, m_{pq} represents the mass flow rate from the phase p to phase q (hydrogen absorption and vaporization). For each cell, the sum of liquid and vapor fraction is the unity (Eq. (22))

$$\alpha_l + \alpha_v = 1 \qquad (22)$$

The mass conservation for each component for both phases is represented by Eqs. (23) and (23), where C_{p_i} is the molar concentration of the component i in the phase p and r_{prod_i} is the production rate of component i by chemical reaction (similarly, r_{cons_i} is the consumption rate of component i by chemical reaction). The chemical reactions occur only in liquid phase.

$$\frac{\partial(\phi \rho_l \alpha_l C_{l_i})}{\partial t} + \nabla.\left(\phi\ \rho_l \alpha_l u_{l_i} C_{l_i}\right) = \phi\ \alpha_l C_{l_i} m_{lv_i} - \phi\ \alpha_l C_{l_i} m_{vl_i} + \phi\ \alpha_l r_{prod_i} - \phi\ \alpha_l r_{cons_i} \qquad (23)$$

$$\frac{\partial(\phi \rho_v \alpha_v C_{v_i})}{\partial t} + \nabla.\left(\phi\ \rho_v \alpha_v u_{v_i} C_{v_i}\right) = \phi\ \alpha_v C_{v_i} m_{vl_i} - \phi\ \alpha_v C_{v_i} m_{lv_i} \qquad (24)$$

The momentum conservation equation for the phase p after averaging is written in Eq. (25). The stress tensor ($\bar{\bar{\tau}}_p$) computes the momentum reduction due the viscous resistance for the mean velocity gradient, the interfacial forces are represented by the term R_{lg} and the momentum variation due mass flux (F_{ij}) between phases is also computed.

$$\frac{\partial(\phi \rho_p \alpha_p u)}{\partial t} = -\phi \rho_p \nabla P + \nabla.\phi \bar{\bar{\tau}}_p + \phi \alpha_p \rho_p g + \sum_{p=1}^{n} F_{ij}\left(u_{ij} - u_{ji}\right) + R_{lg} \qquad (25)$$

The parameter estimation was performed by using the Marquadt-Levenberg method [7], represented in the Eq. (26), minimizing the square of the residues, where ***r*** is the vector of residues, ***J*** is the Jacobian matrix, ***S*** is vector of parameters increments and *C* is a scalar to be adjusted along the convergence process.

$$((\mathrm{J^T J}) + \mathrm{C}.\mathrm{diag}(\mathrm{J^T J}))\mathrm{S} = \mathrm{J^T r} \qquad (26)$$

MODELING RESULTS AND DISCUSSION

The experiments suggest the catalytic bed permeability is not invariant with the Reynolds number. It shows that the flow through this porous medium, for the range of Reynolds number investigated, is not darcinian, once the Reynolds number is higher than 10. Therefore, it was proposed a quadratic relationship between permeability and surface velocity of the fluid for the system under investigation. This relation is shown in Fig. 6. Based on this empirical model, the viscous and inertial resistance coefficients were calculated and fed to the CFD simulation. This permeability, applied to the industrial reactor, resulted in 2.4 % for the average error for the pressure drop prediction.

The values obtained in the laboratory for the bulk density (647.7 kg/m^3/40.43 lb/ft^3) and porosity (0.343) for the catalytic bed are close to the expected values. For confidential reasons, the name of the commercial catalyst manufacturer is not cited.

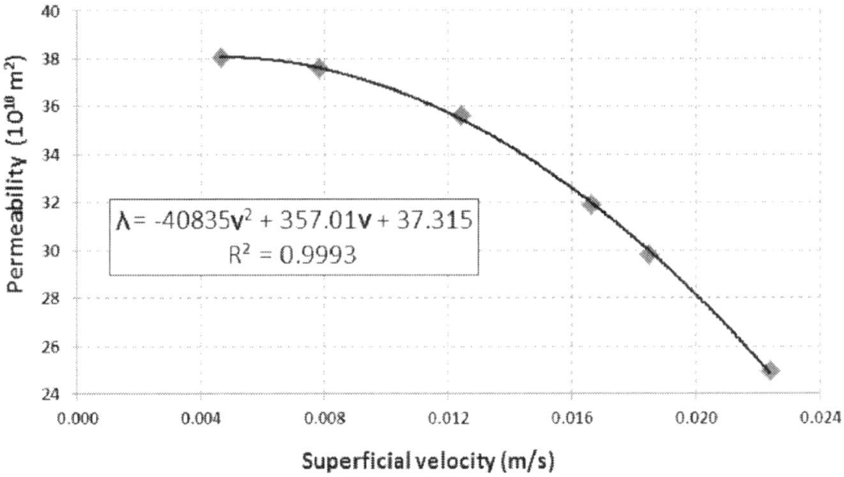

Figure 6. Empirical model to correlate the bed permeability and superficial velocity.

The experimental tests carried out to describe the hydrogen absorption, at equilibrium, by a mixture of hydrocarbons with three carbon atoms (propane, propylene, MA and PD) resulted in the in the Eq. (27), where υ_{H_2} represents the mass fraction of Hydrogen in the liquid phase and T is the absolute temperature in Kelvin.

$$\mathrm{T} = -92{,}386{,}758.1\left(\upsilon_{H_2}\right)^2 - 31{,}136.24\,\upsilon_{H_2} + 322.117 \qquad (27)$$

This is in according to the operational experience, once the industrial catalyst bed temperature does not exceed 49°C (120.2°F). It occurs because the hydrogen concentration in liquid phase is zero when the equilibrium temperature is higher than 49°C (120.2°F); therefore, at this temperature and pressure, there is no exothermic reactions anymore.

After the sensitivity tests, the grid obtained was formed by 138.6 thousand tetrahedral. A grid refinement was performed in order to improve the fluid resolution near the wall. A general view of the grid on wall is shown in Fig. 7.

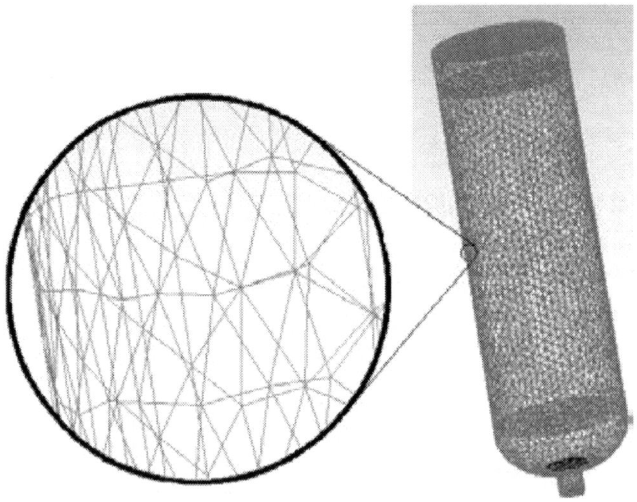

Figure 7. Aspect of the grid on the wall.

The grid quality criteria shown in Table 3 were attended and its values are presented in Table 4. The operational data utilized for the kinetic modelling are inside a typical industrial operational window described in the Table 5; this is a typical operational range for these variables, for this kind of reactor. Once we have used an industrial reactor, we did not have flexibility to increase this range. In fact, this is one of the limitations of this study, the model developed is strictly valid for the ranges presented in the Table 5. However, the advantage of using a phenomenological model (LHHW) is its better representativeness outside of the range where it was developed compared to a non-phenomenological model.

During the catalyst bed campaign, both, MA and PD reactor outlet concentrations increased due the continuous loss of catalyst activity. These data are evidence that the kinetic step is the determinant one to overall MAPD conversion for this reactor in this moment of the catalyst bed campaign, instead of the transport phenomena, once it is not expected the increase of the transport limitations along the catalyst bed campaign.

Table 4. Values of the quality parameters for the obtained grid.

Parameter	Value
Skewness:	<0,50
Aspect ratio of tetrahedral cells:	>0.45
Minimum angle:	>18°
Orthogonal quality:	>0.55

Table 5. Range of the operational data utilized for the kinetic modelling.

Variable	Range of variation
Fresh feed/hydrogen mass flow ratio (kg.h⁻	309 to 367
MA concentration in the fresh feed (%vol):	3.05 to 3.56
PD concentration in the fresh feed (%vol):	2.09 to 2.52
Propane concentration in the fresh feed (%vol):	3.08 to 3.58
Feed temperature (°C/°F):	26 to 28 / 78.8 to 82.4
Outlet pressure (kPa/PSI):	1983 to 2009 / 287.5 to

The final parameter values of the kinetic models are shown in the Table 6. These models were validated by using operational data (separated from the parameters regression operational data) to determine the prediction uncertainness. As mentioned, the MAPD hydrogenation modelling in liquid phase is quite rare in the literature; the few studies found for this system have many simplifications and are based on non-phenomenological models. Therefore, the comparison of the adsorption equilibrium constants identified by this investigation with the few values found in the literature is not useful to shows their consistency. The propane adsorption equilibrium constant was established as zero due it not be significant, this is consistent with the expected strength of adsorption for this component.

Table 6. Final parameter values of the regression of kinetic models.

Parameter	Final value of regression
AK_{MA}	2440.24 kgmol/m^{3}s (152.34 lbmol/ft^{3}s)
EK_{MA}	1180.0 K (1164.3°F)
ZA_{MA}	0.0001993 m^{-1} (0.00006075 ft^{-1})
ZB_{MA}	2.080
ZC_{MA}	313.84
AK_{PD}	1267.68 kgmol/m^{3}s (79.14 lbmol/ft^{3}s)
EK_{PD}	1310.0 K (1898.3°F)
ZA_{PD}	0.0007418 m^{-1} (0.000226 ft^{-1})
ZB_{PD}	1.859
ZC_{PD}	4178.12
AK_{PE}	15.165 kgmol/m^{3}s (0.947 lbmol/ft^{3}s)
EK_{PE}	1610.0 K (2438.3°F)
ZA_{PE}	0.0688989 m^{-1} (0.0210 ft^{-1})
ZB_{PE}	0.946
ZC_{PE}	1376217.0
A_{H2}	18.9 m^3/kgmol (302.75 ft^3/lbmol)
EA_{H2}	5020.0 K (8576.3°F)
A_{MA}	19.0 m^3/kgmol (304.35 ft^3/lbmol)
EA_{MA}	1650.0 K (2510.3°F)
A_{PD}	19.2 m^3/kgmol (307.55 ft^3/lbmol)
EA_{PD}	1380.0 K (2024.3°F)
A_{PE}	1.50 m^3/kgmol (24.03 ft^3/lbmol)
EA_{PE}	450.0 K (350.3°F)
A_{PA}	0.00 m^3/kgmol (0.00 ft3/lbmol)
EA_{PA}	0.00 K (-459.7°F)

An important index for the MAPD selective hydrogenation reactor is the component conversion, defined as the percentage of the inlet quantity of MA or PD hydrogenated in the reactor. The validation results allow us to conclude that the expected MA conversion deviation varies from 0.67 to 2.34% (95% confidence interval), and the expected PD conversion deviation is between 1.50 and 2.60% (95% confidence interval). The expected difference between the measured and the predicted propane outlet concentration, important information to the reactor selectivity analyses, varies from 0.046 to 0.138% (95% confidence interval).

For kinetic models validation purpose, the MA and PD outlet concentration measured in the on-site running reactor and the predicted values by the models developed in this work, in the same process conditions, are compared in the Fig. 8. The data utilized in this comparison were not applied to the parameters regression algorithm. The equivalent comparison for the propane is shown in Fig. 9.

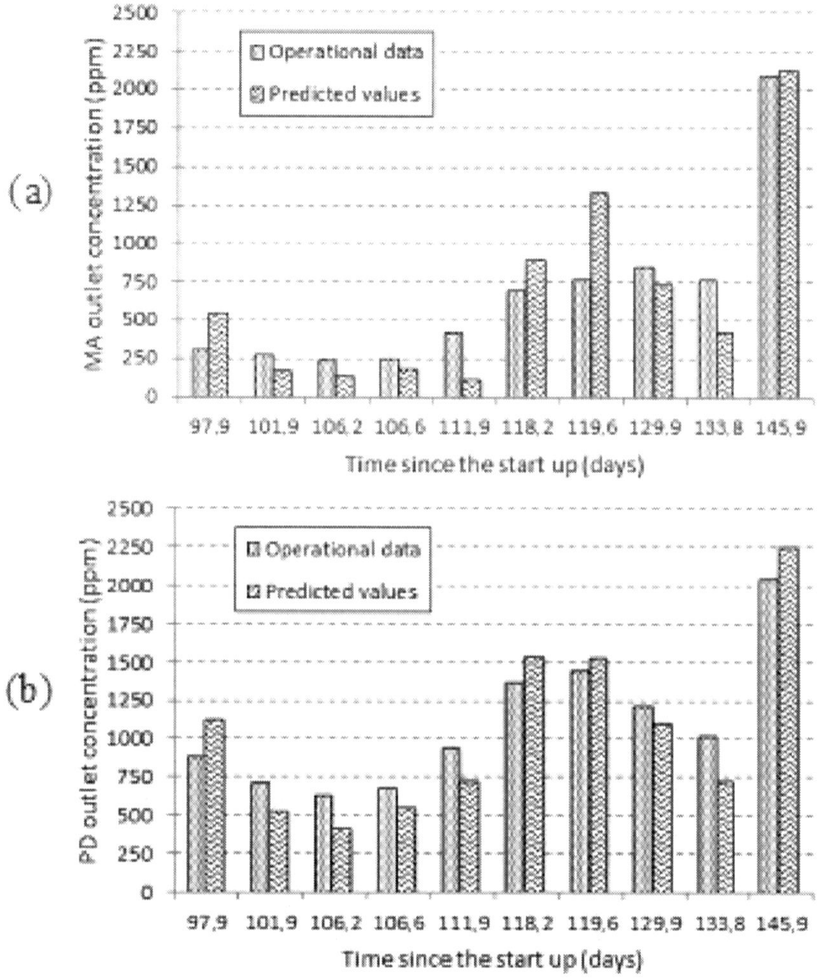

Figure 8. Comparison between the (a) MA and (b) PD outlet concentrations (measures versus predicted), in the model validation experiments.

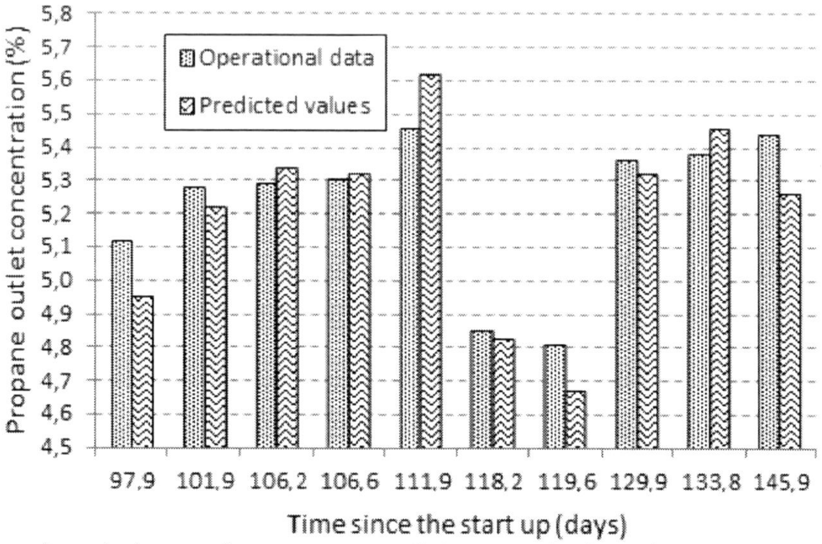

Figure 9. Comparison between the propane outlet concentrations (measures versus predicted), in the model validation experiments.

The same models validation analysis shows in the Fig 10 the good agreement between the predicted and the measured axial temperature profile of the catalytic bed.

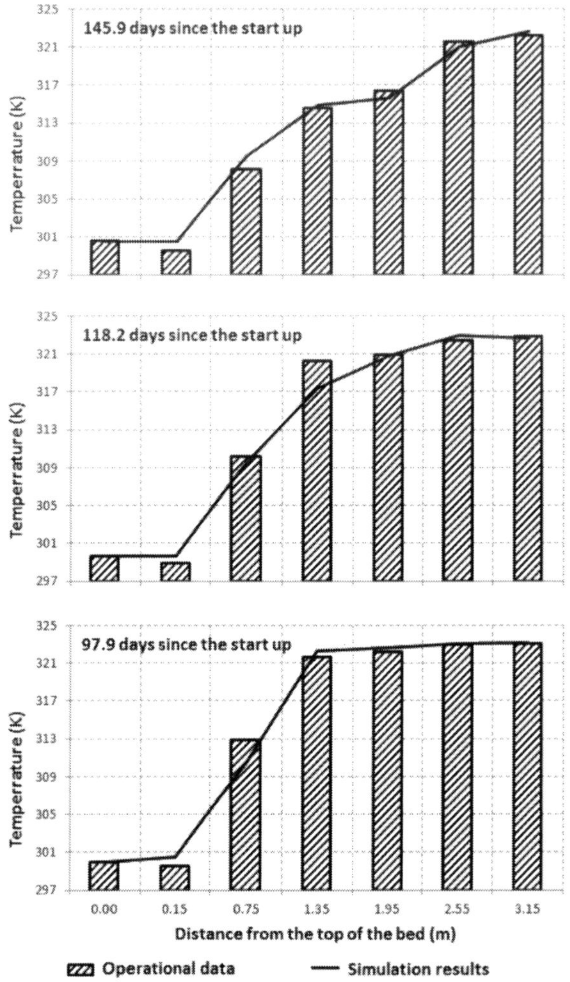

Figure 10. Comparison between the axial temperature profile (measures versus predicted), in the model validation experiments.

Fluid dynamic aspects

As represented in the Fig. 11 (referred to a catalytic bed running for 118.2 days), the axial velocity varies from zero, close to wall, to 0.20 m/s (0.66 ft/s), the maximum velocity being a function of the height of the cross section. This behavior is similar at various heights of the bed and days of campaign. These results show that the value of N (vessel internal diameter divided by catalytic particle diameter, in this system, around 400) is high enough to prevent a poor flow distribution.

Despite the temperature elevation, due the exothermic reactions, and consequently increase of the vaporization rate, the hydrogen consumption increases the liquid phase volume fraction. This effect is shown in Fig. 12, referred to a catalytic bed running for 118.2 days. Based on the simulation results, the vapor phase volume fraction decreases, in this example, from 0.53 to 0.41. These data show the importance of CFD to represent correctly the reactor behavior and to isolate the effect of the kinetics.

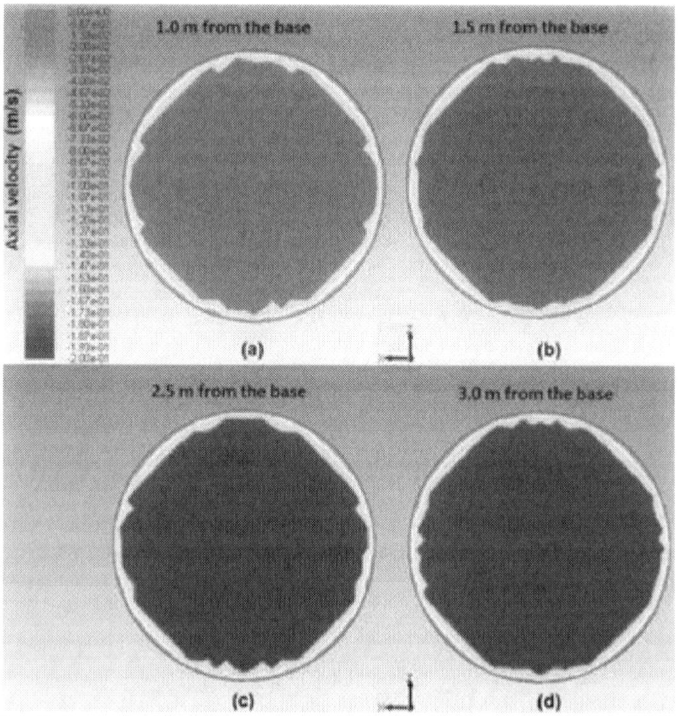

Figure 11. Radial contours of axial velocity at various heights of the catalytic bed.

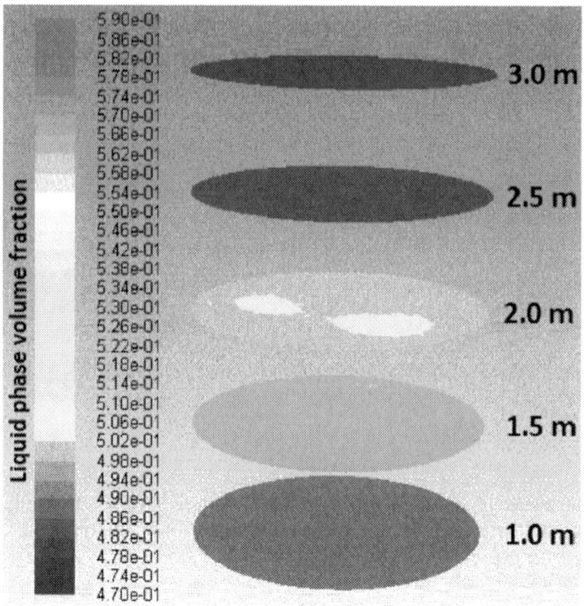

Figure 12. Axial and radial liquid phase volume fraction profile.

The reaction rates distribution is illustrated in Fig. 13. This figure represents the axial central cross section of the reactor, after 118.2 days since the start up. The synergic effect of concentrations and chemical characteristics results in an initial higher hydrogenation rate of MA, compared to the PD hydrogenation rate. The propylene hydrogenation rate is even lower than the PD hydrogenation rate and occurs posteriorly, once it needs high temperature, upper to 310 K (98.3°F), and low MAPD concentration (to reduce the competition).

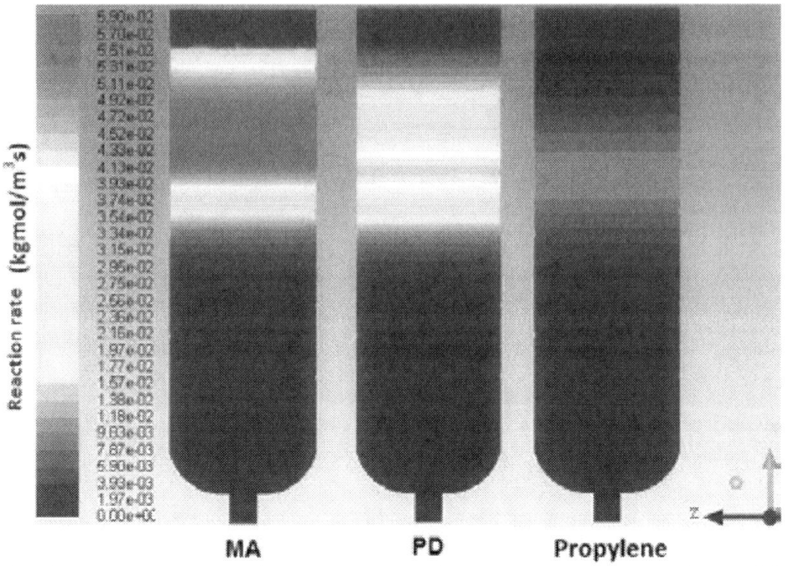

Figure 13. Reaction rate contours in the axial central cross section of the reactor.

The comparison between the hydrogenation rates of MA and PD is expressed by the ratio of these rates, as described by Eq. (28).

$$\frac{r_{MA}}{r_{PD}} = \frac{k_{MA}\,K_{MA}[MA]\,Act_{MA}}{k_{PD}\,K_{PD}[PD]\,Act_{PD}} \tag{28}$$

$$\frac{r_{MA}}{r_{PD}} = \frac{AK_{MA}\,A_{MA}[MA]\exp\left(-EK_{MA}\left(\frac{1}{T}-\frac{1}{310}\right)\right)\exp\left(EA_{MA}\left(\frac{1}{T}-\frac{1}{310}\right)\right)\,(1/(1+\frac{\exp\left(ZA_{MA}.h.t^{ZB_{MA}}\right)}{ZC_{MA}}))}{AK_{PD}\,A[PD]\exp\left(-EK_{PD}\left(\frac{1}{T}-\frac{1}{310}\right)\right)\exp\left(EA_{PD}\left(\frac{1}{T}-\frac{1}{310}\right)\right)\,(1/(1+\frac{\exp\left(ZA_{PD}.h.t^{ZB_{PD}}\right)}{ZC_{PD}}))} \tag{29}$$

$$\frac{r_{MA}}{r_{PD}} = \frac{AK_{MA}\,A_{MA}\,[MA]\,(1+\frac{\exp\left(ZA_{PD}.h.t^{ZB_{PD}}\right)}{ZC_{PD}})}{AK_{PD}\,A_{PD}\,[PD]\,(1+\frac{\exp\left(ZA_{MA}.h.t^{ZB_{MA}}\right)}{ZC_{MA}})}\exp\left((EA_{MA}-EK_{MA}-EA_{PD}+EK_{PD})\left(\frac{1}{T}-\frac{1}{310}\right)\right) \tag{30}$$

After replacing the values of these parameters, we reach the Eq. (31).

$$\frac{r_{MA}}{r_{PD}} = \frac{[MA]}{[PD]}\exp\left(\frac{400}{T}-3.23462\right)\frac{(4178.12+\exp(0.0007418.h.t^{1.859}))}{(313.84+\exp(0.0001993.h.t^{2.080}))} \tag{31}$$

Therefore, the ratio of MA and PD hydrogenation rate in the bottom layer of catalyst bed (h = 0), where the catalyst has full activity, is described by the Eq. (32). Based on this equation, we concluded the ratio of MA and PD hydrogenation rate, for the same concentration of both components, typical operational conditions and same catalyst activity, is close to 1.8. This result is coherent with the high reactivity of MA, compared to the PD.

$$\frac{r_{MA}}{r_{PD}} = \frac{[MA]}{[PD]}\exp\left(\frac{400}{T}-0.64589\right) \tag{32}$$

Reactor Optimization

The response surface, referred to the MAPD output, adjusted from the simulation data, is represented graphically in Fig. 14. This surface shows the direction of MAPD outlet concentration evolution when the reactor recirculation flow rate and the hydrogen flow rate are modified. This kind of graph is useful to show the combined effect of various variables on the reactor output parameters.[39] It is important to remember that this particular evaluation was performed considering a certain fresh feed flow rate, temperature and MAPD concentration in the fresh feed. In this case, the catalytic bed had run for 125 days. The recirculation flow ranging from 40000 to 48000 kg/h (88185 to 105822 lb/h) and hydrogen flow ranging from 95 to 125 kg/h (209.4 to 275.6 lb/h). This response surface allows us to realize that, for these process conditions, the MAPD conversion increases proportionally to the recirculation flow rates and hydrogen flow fed to the reactor. This behavior can be explained by the cooling effect of the recirculation, removing heat from the system. Therefore, when the recirculation flow rate increases, it will increase also the hydrogen concentration in the liquid phase. This graphic also allows us to conclude the MAPD conversion is much more sensitive to the recirculation than to the hydrogen feed rate.

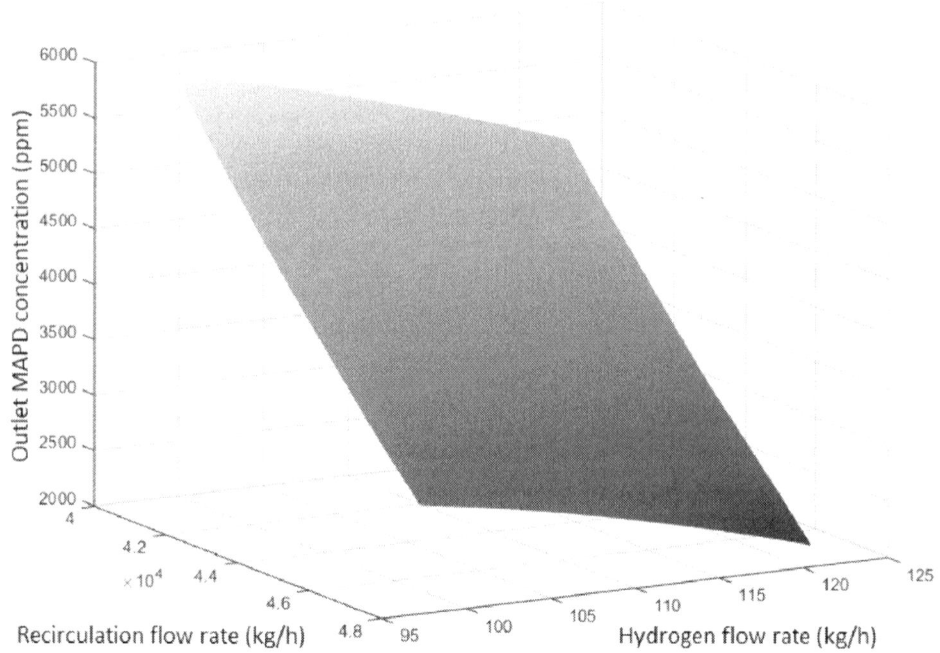

Figure 14. Response surface for MAPD outlet concentration as function of recirculation flow rate and hydrogen flow rate.

Another key performance parameter of the catalytic bed is its selectivity, the capability to hydrogenate MAPD instead propylene.[38] The response surface of this reactor for the amount of propylene hydrogenated, related to the recirculation flow rate and the hydrogen flow rate, is a mean to express its selectivity. This surface is shown in Fig. 15, obtained with simulation data under the same operational conditions used to generate Fig. 14.

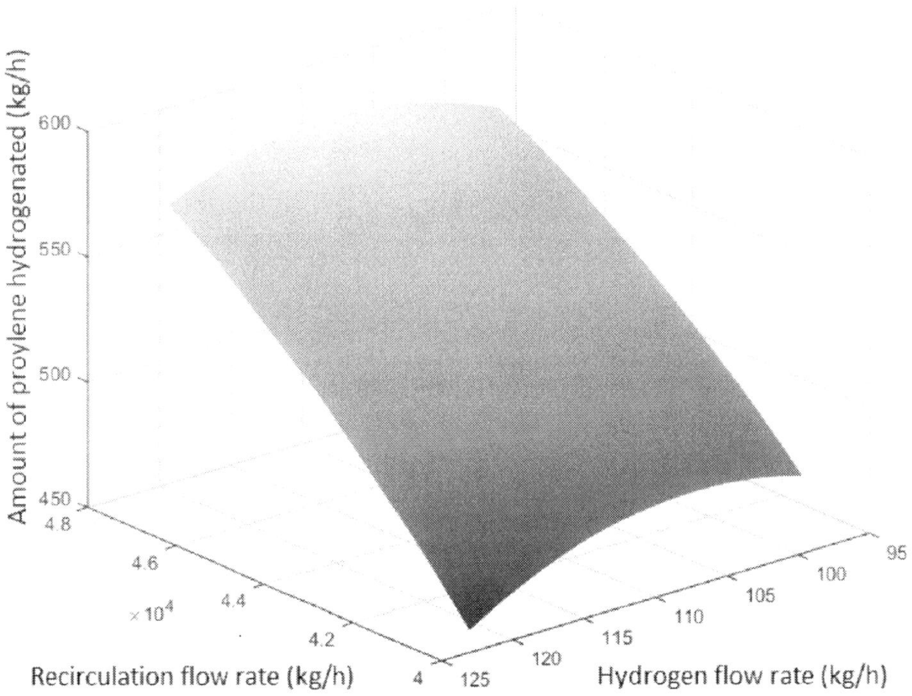

Figure 15. Response surface for the reactor selectivity (amount of propylene hydrogenated) as function of recirculation flow rate and hydrogen flow rate.

The behavior of the reactor selectivity, related to the recirculation flow variations is opposite to the observed effect to catalytic bed activity. The catalytic bed selectivity increases when the recirculation flow rate is reduced. This behavior can be explained by the effect of the MAPD concentration on the probability of MAPD hydrogenation, instead of propylene. A small influence of hydrogen flow rate on the reactor selectivity was observed.

CONCLUSIONS

A phenomenological kinetic model for the MAPD hydrogenation reaction in liquid phase, based on LHHW mechanism, was determined experimentally by using data from an on-site running industrial TBR, including transport aspects and catalyst deactivation, with CFD assistance. This was a realistic model, once many aspects related to trickle regime, porous media, double phase flowing and mass transfer between phases were taken in account. This way, this work fills an important gap in the literature, once studies regarding this reaction, considering both the kinetic and transport aspects, are quite rare. The model validation shows the expected absolute difference between the measured and predicted MA conversion is in the range from 0.67 to 2.34% (95% confidence interval) and between 1.50 and 2.60% (95% confidence interval) for PD, while the expected difference between the measured and predicted propane outlet concentration is in the range from 0.046 to 0.138% (95% confidence interval). A response surface based on the obtained models shows the MAPD conversion increases with the recirculation ratio, on the other hand, the same action reduces the reactor selectivity, while the effect of hydrogen feed rate on the MAPD conversion and selectivity is not relevant. The obtained models allow us to understand the phenomena inside the TBRs applied to MAPD hydrogenation and to design new TBRs to this function.

LIABILITY WARNING

All recommendations in this article are made without guarantee on the part of the author, therefore the author disclaim any liability in connection with the use of this information.

ACKNOWLEDGEMENTS

The authors are grateful to Braskem (Brazilian petrochemical company) and to Faculdade de Tecnologia SENAI CIMATEC for the support to this investigation.

REFERENCES

(1) Plotkin, J., 2015. Understanding the Global Petrochemical Industry: Propylene [WWW Document]. IHS Chem. Week. URL http://www.chemweek.com/lab/Understanding-the-Global-Petrochemical-Industry-Propylene_70586.html (accessed 4.12.16).

(2) Akah, A., Al-Ghrami, M., 2015. Maximizing propylene production via FCC technology. Appl. Petrochemical Res. 5, 377–392. doi: 10.1007/s13203-015-0104-3.

(3) Jeffrey J. Siirola, 2014. Modeling and simulation of the polymeric nanocapsule formation process. AIChE J. 60, 810–819. doi: 10.1002/aic.14368.

(4) Wu, W., Li, Y.-L., Chen, W.-S., Lai, C.-C., 2011. Kinetic Studies and Operating Strategies for an Industrial Selective Hydrogenation Process. Ind. Eng. Chem. Res. 50, 1264–1271. doi: 10.1021/ie101183c.

(5) Falqi, F., 2009. The Miracle of Petrochemicals Olefins Industry, 1th ed. Universal-Publishers, Florida USA.

(6) Wang, B., Froment, G.F., 2005. Kinetic Modeling and Simulation of the Selective Hydrogenation of the C 3 -Cut of a Thermal Cracking Unit. Ind. Eng. Chem. Res. 44, 9860–9867. doi: 10.1021/ie050891p.

(7) Froment, G.F., Bischoff, K.B., 1990. Chemical Reactor Analysis and Design, 2nd Editio. ed. John Wiley & Sons, New York.

(8) Dudukovic, M.P., 2007. Relevance of Multiphase Reaction Engineering to Modern Technological Challenges. Ind. Eng. Chem. Res. 46, 8674–8686. doi: 10.1021/ie070371p.

(9) Tohidian, T., Dehghani, O., Rahimpour, M.R., 2015. Modeling and simulation of an industrial three phase trickle bed reactor responsible for the hydrogenation of 1,3-butadiene: A case study. Chem. Eng. J. 275, 391–404. doi: 10.1016/j.cej.2015.03.035.

(10) Ranade, Vivek V.; Chaudhari, Raghunath V.; Gunjal, P.R., 2011. Trickle Bed Reactors. Elsevier.

(11) Samimi, F., Modarresi, Z.K., Dehghani, O., Rahimpour, M.R., Bolhasani, A., 2015. Application of response surface methodology for optimization of an industrial methylacetylene and propadiene hydrogenation reactor. J. Taiwan Inst. Chem. Eng. 46, 51–64. doi: 10.1016/j.jtice.2014.09.003.

(12) Ancheyta, J., 2011. Modelling and Simulation of Catalytic Reactors for Petroleum Refining, Focus on Catalysts. John Wiley & Sons, New Jersey.

(13) Uygur, H., Atalay, S., Savasci, T.O., 1998. Kinetics of Liquid Phase Selective Hydrogenation of Methylacethylene and Propadiene in C3 Streams. J. Chem. Eng. Japan 31, 178–186.

(14) Yu, Z., Gao, B., Xu, J., Zhang, J., 2002. Kinetic study of C3 cut liquid-phase selective hydrogenation over supported metal catalyst. Ind. Catal. 10, 11–16.

(15) Atta, A., Roy, S., Nigam, K.D.P., 2007. Prediction of pressure drop and liquid holdup in trickle bed reactor using relative permeability concept in CFD. Chem. Eng. Sci. 62, 5870–5879. doi: 10.1016/j.ces.2007.06.008.

(16) Bozzano, G., Dente, M., Manenti, F., Milano, P., Chimica, D., Chimica, I., Natta, G., Leonardo, P., Corna, P., Masserdotti, F., 2014. Fluid Distribution in Packed Beds . Part 1 . Literature and Technology Overview. Ind. Eng. Chem. Res. doi: 10.1021/ie402137z.

(17) Mavriplis, C., 2012. Interdisciplinary CFD. Int. J. Comut. Fluid Dyn. 26, 333–335. doi: 10.1080/10618562.2012.739353.

(18) Prades, L., Dorado, A.D., Climent, J., Guimerà, X., Chiva, S., Gamisans, X., 2017. CFD modeling of a fixed-bed biofilm reactor coupling hydrodynamics and biokinetics. Chem. Eng. J. 313, 680–692. doi: 10.1016/j.cej.2016.12.1070.

(19) Lopes, R.J.G., Quinta-Ferreira, R.M., 2009. CFD modelling of multiphase flow distribution in trickle beds. Chem. Eng. J. 147, 342–355. doi: 10.1016/j.cej.2008.11.048.

(20) Wang, Y., Chen, J., Larachi, F., 2013. Modelling and simulation of trickle-bed reactors using computational fluid dynamics: A state-of-the-art review. Can. J. Chem. Eng. 91, 136–180. doi: 10.1002/cjce.20702.

(21) Gunjal, P.R., Ranade, V. V., 2007. Modeling of laboratory and commercial scale hydro-processing reactors using CFD. Chem. Eng. Sci. 62, 5512–5526. doi: 10.1016/j.ces.2007.01.078.

(22) Dixon, A.G., Nijemeisland, M., 2001. CFD as a Design Tool for Fixed-Bed Reactors. Ind. Eng. Chem. Res. 40, 5246–5254. doi: 10.1021/ie001035a.

(23) Zikanov, O., 2010. Essential computational fluid dynamics, 1st ed. John Wiley & Sons, New Jersey.

(24) Patankar, S. V., 1980. Numerical heat transfer and fluid flow, 1st ed, International Journal of Heat and Fluid Flow. Hemisphere Publishing Corporation, New York.

(25) Yu, H., Mao, Z., Dai, W., Peng, J., Zhai, M., Wei, G., 2012. Highly Selective Pd/Al2O3 Catalyst for Hydrogenation of Methylacetylene and Propadiene in Propylene Stream Prepared by Y Radiation. Appl. Catal. A Gen. 445–446, 246–251. doi: 10.1016/j.apcata.2012.08.036.

(26) Singh, U.K., Vannice, M.A., 2001. Kinetics of liquid-phase hydrogenation reactions over supported metal catalysts A review. Appl. Catal. A Gen. 213, 1–24.

(27) Delgado, J.M.P.Q., 2005. A critical review of dispersion in packed beds. Heat Mass Transf. 42, 279–310. doi: 10.1007/s00231-005-0019-0.

(28) Schwidder, S., Schnitzlein, K., 2012. A new model for the design and analysis of trickle bed reactors. Chem. Eng. J. 207–208, 758–765. doi: 10.1016/j.cej.2012.07.054.

(29) Ansys, 2014. Ansys Theory Guide.

(30) Eisfeld, B., Schnitzlein, K., 2001. The influence of confining walls on the pressure drop in packed beds. Chem. Eng. Sci. 56, 4321–4329.

(31) Bear, J., 1972. Dynamics of Fluids in Porous Media, American Elsevier publishing company. American Elsevier Publishing Company, New York.

(32) Guo, B., Yu, A., Wright, B., Zulli, P., 2006. Simulation of Turbulent Flow in a Packed Bed. Chem. Eng. Technol. 29, 596–603. doi: 10.1002/ceat.200500292.

(33) Chávez, L.M., Alonso, F., Ancheyta, J., 2014. Vapor–liquid equilibrium of hydrogen–hydrocarbon systems and its effects on hydroprocessing reactors. Fuel 138, 156–175. doi: 10.1016/j.fuel.2014.03.062.

(34) Ramdharee, S., Muzenda, E., Belaid, M., 2013. A review of the equations of state and their applicability in phase equilibrium modeling, in: Internation Conference on Chemical and Environmental Engineering (ICCEE 2013). pp. 84–87.

(35) Raeini, A.Q., Blunt, M.J., Bijeljic, B., 2012. Modelling two-phase flow in porous media at the pore scale using the volume-of-fluid method. J. Comput. Phys. 231, 5653–5668. doi: 10.1016/j.jcp.2012.04.011.

(36) Lopes, R.J.G., Quinta-Ferreira, R.M., 2009. Turbulence modelling of multiphase flow in high-pressure trickle-bed reactors. Chem. Eng. Sci. 64, 1806–1819. doi: 10.1016/j.ces.2008.12.026.

(37) Hirsch, C., 2007. Numerical computation of internal and external flows. Volume 2, 2nd ed, John Wiley and Sons, Chichester. Elsevier.

(38) Mohundro, E.L., 2003. Overview on C2 and C3 Selective Hydrogenation in Ethylene Plants, in: AIChE Spring Meeting. New Orleans.

(39) von Aretin, T., Standl, S., Tonigold, M., Hinrichsen, O., 2017. Optimization of the product spectrum for 1-pentene cracking on ZSM-5 using single-event methodology. Part 2: Recycle reactor. Chem. Eng. J. 309, 873–885. doi: 10.1016/j.cej.2016.10.103.

Molecular composition modelling of Vacuum Gas Oils via a Molecular Reconstruction approach

E. Gomez Suarez[1], L. P. de Oliveira*,[1], J. J. Verstraete[1]

[1] IFP Energies nouvelles, Rond-point de l'échangeur de Solaize - BP 3, 69360 Solaize, France

Keywords: Composition modelling, Molecular Reconstruction, Monte-Carlo method, Information entropy, Vacuum Gas Oils

A Vacuum Gas Oil (VGO) is a complex hydrocarbon mixture of several thousand different chemical species. Describing the chemical composition of the feedstocks is essential to the design and operation of these refining processes. Although the most advanced analytical techniques, such FT-ICR/MS, enable to identify a large number of compounds and classes of chemical families, a molecular composition of the VGO cuts cannot be obtained directly, even by means of cutting-edge analytical techniques. To overcome this drawback, a molecular representation of the feedstock can be built using a molecular reconstruction algorithm. The basic idea behind this type of algorithms is to generate a discrete set of molecules from overall petroleum analyses, model hypotheses and chemical knowledge.

During previous work [1-3], a two-step molecular reconstruction algorithm was developed to model the composition of VGO cuts by using two methods in series: Stochastic Reconstruction (SR) [4] and Reconstruction by Entropy Maximization (REM) [5]. The SR step uses a Monte-Carlo approach to generate the initial set of molecules. In the REM, the mole fractions of the molecules are adjusted by maximizing an information entropy criterion.

Even that this algorithm can successfully reconstruct the VGO cuts, the SR step still remains a computationally demanding algorithm.

This work focuses on the development of an alternative approach of this base algorithm which allows reducing the computational burden when several VGO cuts need to be reconstructed. The idea of this alternative approach, called indirect two-step molecular reconstruction (SR-REM) algorithm, consists of using the SR method to build a reference database of molecules for all VGO cuts. The abundance of each molecule of the database is then adjusted via the REM method in order to obtain a synthetic mixture whose properties are closer to the analytical data of the VGO to be represented.

The indirect SR-REM algorithm was applied to eight different VGO cuts that are either straight run VGOs obtained from crude oils from different geographical origins or conversion VGOs from various refining processes. The VGOs have been characterized by means of elementary analysis (C,H, S, N), density, detailed Fisher mass spectrometry, liquid chromatography S.A.R. distribution as well as sulfur and nitrogen (total, basic and non-basic nitrogen) simulated distillation.

The reference database was built by collecting the mixtures of 5000 molecules for three VGO cuts. In this work, VGO molecules are constructed from 16 structural attributes which enable to generate acyclic molecules (paraffins, iso-paraffins) and cyclic molecules, the latter being composed of napthenic rings, aromatic rings and heterocyclic rings (thiophenes, pyrolles and pyridines). The structural attributes are described by three type of PDFs: Histograms, Exponential distributions and Gamma distributions.

The application of the indirect molecular reconstruction algorithm yields a synthetic mixture of 15 000 molecules that mimics very well the properties of all VGOs. The results of the indirect reconstruction are similar with those obtained with the direct molecular reconstruction algorithm. This molecular reconstruction algorithm advantageously combines the power of the SR method to generate appropriate molecules with the flexibility of the REM method, while completely eliminating the computational burden when several VGOs need to be reconstructed.

References

[1] J. J. Verstraete, N. Revellin.,H. Dulot, D. Hudebine, *Preprints Paper - ACS, Division Fuel Chemistry*. **49** (2004) 20-21

[2] N. Charon-Revellin, H. Dulot,C. López-García, J. Jose, *Oil & Gas Science and Technology*. **66** (2011) 479-490

[3] A. Alavarez-Majmutov, R. Gieleciak, J. Chen, *Enery & Fuels*. **29** (2015) 7931-7940

[4] M. Neurock , A. Nigam, D.M. Trauth, M.T. Klein, *Chem. Eng. Sci.*, **49** (24) 4153-4177

[5] D. Hudebine, J.J. Verstraete *Chem. Eng. Sci.*, **59** (2004) 4755

550568 Fault Isolation Using a Multiscale PCA Reconstruction-Based Approach

Process monitoring and control have become more automated in the past decade through the utilization of data-driven techniques. This is motivated by the need to improve the efficiency of operation and to avoid undesirable outcomes that may result from failing sensors. Thus, effective fault detection and isolation methods are essential for efficient operation of industrial processes. This work proposes a new fault isolation technique for process monitoring applications, which extends the abilities of PCA reconstruction-based isolation using wavelet multiresolution (MR) analysis.

Multiresolution wavelet analysis is a well-established and widely-implemented tool for fault detection. The wavelet coefficients inherently possess characteristics that enable efficient separation of deterministic and stochastic features in measured data, and help deal with autocorrelated noise [1]. PCA reconstruction-based isolation has also shown advantages over other PCA isolation techniques, such as structured residuals and contribution plots [2]. This is due to its ability to deal with the residual smearing effects, which hinder effective fault isolation.

The main contribution of this work is the development of an algorithm for isolation that utilizes the benefits of both MR analysis and PCA reconstruction-based fault isolation. The effectiveness of the developed algorithm will be demonstrated through a simulated example of a sensor fault that occurs in a chemical process unit, e.g. a CSTR reactor. The isolation results obtained by implementing PCA reconstruction-based contribution with MR analysis is compared to those obtained by implementing only the PCA reconstruction-based method (without MR analysis) in order to examine the advantages achieved by the developed algorithm. Further validation of the results will be carried out, by comparing the developed algorithm to previous work that performed MR analysis with contribution plot isolation [3].

References

[1] B. Bakshi, Multiscale PCA with application to multivariate statistical process monitoring, AIChE J. 44 (1998) 1596–1610.

[2] R. Dunia, F. Systems, S.J. Qin, T.F. Edgar, T.J. Mcavoy, Identification of faulty sensors using principal component analysis, AIChE J. 42 (1996).

[3] M. Misra, H.H. Yue, S.J. Qin, C. Ling, Multivariate process monitoring and fault diagnosis by multi-scale PCA, Comput. Chem. Eng. 26 (2002) 1281–1293.

19th AiChe Spring Meeting, 31st March - 4th April 2019, New Orleans, Louisiana (US)

Fouling early detection in ethylene plant Steam Turbines

Maddalena Pondini[a], Giuseppe Salerno[a], Giancarlo Gaudiuso[a]

[a]© 2018 BAKER HUGHES, a GE company, Florence, Italy

Abstract

The petrochemical industry is facing more and more stringent requirements in terms of process reliability and maintenance costs. Thus, companies are in the need for advanced technologies to monitor the operation of critical process equipment.

The phenomenon of fouling builds up gradually with operation and represents one of the major issues affecting turbomachinery in charge gas and refrigerant trains. Steam Turbines fouling, in particular, is caused by poor steam purity and typically results in a reduced steam swallowing capacity due to the increase of pressure in machine section at constant steam flow.

To enable an early detection of this phenomenon the variation of the Stodola constant (K_{ST}) calculated across machine wheel chamber and exhaust sections is compared with K_{ST} resulting from the machine design tool calculation at same steam conditions. The long-term monitoring of this parameter can be applied to prevent severe machine damage and to suggest the timing for machine cleaning procedures.

The results of this comparison on a specific use-case will be presented in detail in this paper.

1. Introduction

The increasing pressure on the economics in petrochemical industry is forcing it to face more and more stringent requirements in terms of process reliability and maintenance costs. The achievement of these targets requires both a closer look at critical equipment integrity and the possibility of pushing operation towards less frequent turnarounds. In this environment, one of the main enablers is the deep digital transformation that is currently taking place in the different sectors of the industry, which, by means of data analysis techniques will open the way to a wide range of opportunities still unexplored [1].

Turbomachinery equipment is amongst the most critical components affecting process yield and maintenance requirements, thus, an effective way of increasing their reliability is to translate manufactures know-how into valuable analytics and algorithms that process site data. The present study aims at describing a new methodology for the early detection of fouling issues in steam turbines. This phenomenon is caused by impurities in the steam and severely affects components with operation by impacting the performances and, consequently, the efficiency of the whole process. In order to help plant managers preventing unexpected downtimes and the increasing degradation of process yields the main countermeasure is to install online washing systems which let saturated steam pass through the steam path and dissolve the soluble deposits. In addition, it is also necessary to integrate this system with dedicated strategies for the monitoring of fouling build-up, which normally reflects in a poor steam swallowing capacity through the steam turbine and, consequently, pressures increase in the wheel chamber section.

The paper is organized as follows: Sec. 2 provides an overview on the causes and main countermeasures against fouling, Sec. 3 presents the methodology proposed for the early detection of the phenomenon; Sec. 4 depicts the results of a long-term monitoring of steam turbines data on a specific use-case. Finally, Sec. 5 provides some concluding remarks and hints for future work.

2. The fouling phenomenon in Mechanical Drive (MD) steam turbines

Fouling is a major problem for steam turbine operability, resulting in performance deterioration, increased thrust and risk of stress corrosion cracking; this depends on the contaminants dispersed in the steam. The build-up of deposits in turbines due to impurities dispersed in the steam can cause both thermodynamic and mechanical problems. From a thermodynamic point of view, deposits decrease efficiency and may also result in a reduced maximum inlet steam intake capacity. Corrosion, on the other hand, may result in mechanical problems, including equipment failure, with detrimental effects on machine reliability and availability, in particular circumstances (e.g. when chlorides or sulphates are present). It is therefore of fundamental importance to deeply understand the causes and the mechanisms affecting the chemical quality of the steam, as well as to develop and validate good practices for the monitoring of this type of issues.

MD steam turbines operating in petrochemical and refinery processes are traditionally affected by fouling due to their key role in ensuring a long-term operation to achieve high process yields with low maintenance costs. On the other hand, pushing operation forward not only means uptime but also translates into machine performances deterioration which determine a poorer driving capability of the whole compression train [2].

This happens due to the deposition occurring on steam turbine blades and vanes taking place when the concentration of any contaminant exceeds the solubility limit in the superheated steam at the local pressure and temperature. In such conditions the impurity precipitates and starts to deposit on turbine surfaces in solid phase. As a result, the geometry of the steam path comprised between wheel chamber and exhaust sections changes and determines a lower steam intake capacity of the machine. This brings the steam turbine control valve to operate closer to its maximum opening and eventually not being able of maintaining the desired operational speed anymore. The typical compounds that are responsible of machine fouling are silica compounds like SiO_2, which can also combine with sodium compounds (namely NaOH or NaCl), if present, and form sodium silicates $Na_2Si_2O_5$. The latter compound is soluble in water and can be washed off, its deposition pattern, together with the ones of sodium compounds, are shown in Figure 1 below.

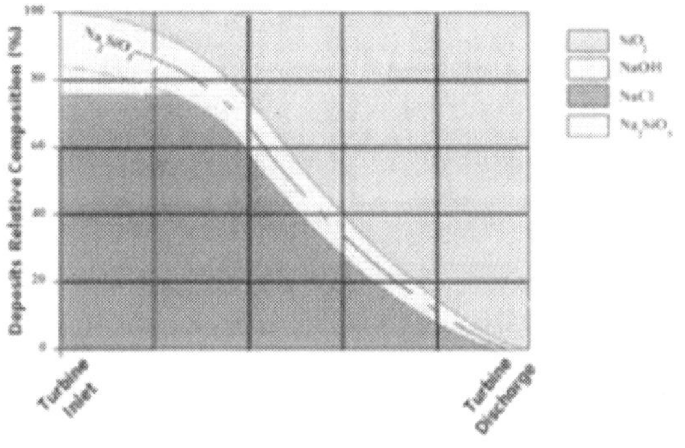

Figure 1 - Typical deposition patterns of sodium and silica compounds

In order to re-establish acceptable operating conditions cleaning operations are needed. Different techniques are possible, mechanical cleaning ones, which require machine disassembly, and hot water cleaning with machine in slow roll [2], both options come along with the need of shutting down the process which is time consuming and, thus, generating revenue losses for the customer. Online washing systems represent the most suitable solution to wash off soluble compounds from the steam turbine by increasing the amount of wetness inside the turbine while the production process is on [2]. Together with this effective countermeasure, monitoring strategies are needed to enable a direct comparison between machine performances measured and the ones expected if the steam turbine was operating as a new one. In the following paragraph a new methodology for the early detection of deposits build-up is described and the results of its implementation are shown for a specific use-case under analysis.

3. Early detection methodology

As aforementioned above, the deposits occurring in the steam path determine a reduction of the passing section and, consequently, a change in the geometry of the machine. This results in an increase of the pressure at the upstream section, normally the wheel chamber section, to guarantee a constant steam flow. Due to this correlation between machine fouling, wheel chamber pressure and steam flow a suitable monitoring strategy should start from the analysis of the behaviour of such parameters. The law that is most commonly used to describe the dependency of steam flow on its conditions and on machine geometry is the Stodola law [4], which in its simplified form can be synthesized as per equation below:

$$\dot{m} = k_{ST}\sqrt{\frac{p_{in}^2 - p_{out}^2}{T_{in}}}$$

where k represents the Stodola constant. When considering the machine sections comprised between wheel chamber and exhaust (p_{in} = p_wheel_chamber, p_{out} = p_exhaust) in 'new and clean' machine state, this parameter is constant as a demonstration of the unvaried geometry. This last condition is a requirement for a representative description of the relationship between the parameters through the formula. Thus, for cases in which deposition is occurring and unvaried geometry condition is no longer valid, researchers have approached the problem by focusing on regression methods based on data measured at site. In [4] a machine learning technique called Support Vector Regression is proposed to predict mass flow in machine fouled conditions. This method shows accuracy in the prediction, but has the drawback of a poor extrapolation capability if the machine is operated out of the range used for model training. Additionally, it can't be applied 'as is' for any type of steam turbine, but it would need new datasets for model training and parameters identification. Similar considerations can be given for the approach adopted in [5], where regression methods are applied to fouling fault data.

The methodology presented in this paper is based on a comparison between expected steam turbine performances and measured ones. To enable this type of approach the know-how of the manufacturer, namely the machine design tool, can be applied to run operating points according to the actual operating conditions at site. Figure 2 below depicts the layout of the methodology applied, where $p_{expected}$ is the wheel chamber pressure determined as a function of the geometry of the steam path, the steam conditions and the mass flow.

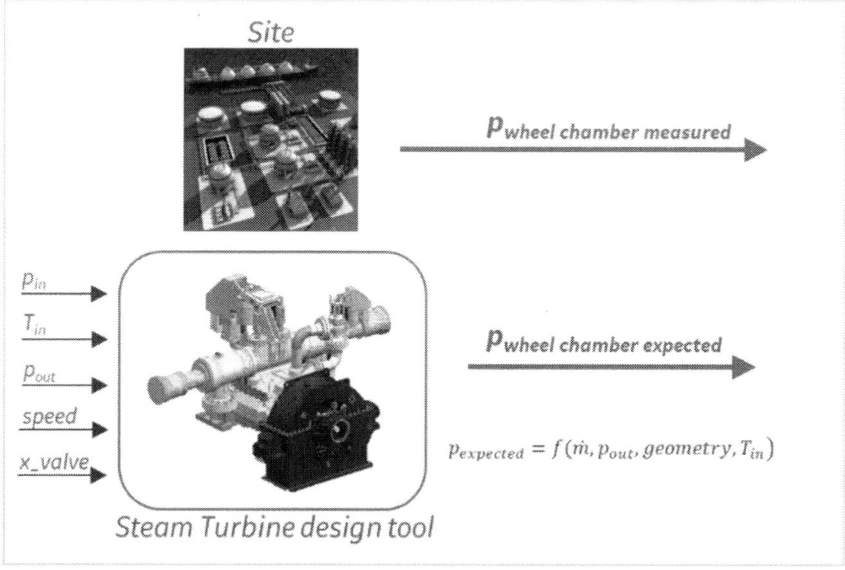

Figure 2 - Methodology applied

The comparison between the measured wheel chamber pressure and the expected one enables an early detection of fouling build-up. Section 4 describes the results of such comparison conducted on a specific case study where wheel chamber pressure sensor was installed for monitoring purpose.

4. Case study

The case-study proposed is the result of a one-year data collection and analysis campaign conducted on a hydrogen recycle compression train operating in a refinery. Customer required BHGE's expertise on compressor and steam turbine design to perform a long-term analysis of site data and to retrieve useful information regarding machine health status and operation. The aim of the study was to gather valuable information from site data in terms of machine performance degradation and maintenance needs. To achieve this, dedicated machine analytics have been implemented which enable an early detection of wrong operation and, possibly, prevent severe damages and downtime. One of the issues that have been targeted is the fouling of the steam turbine which the customer identified as one of the main obstacles to achieve the expected machine performances in the past. Thus, wheel chamber pressure sensor and control valve command have been connected to gather valuable data for a coherent comparison between expected performances (output of the design tool) and actual ones. The results of this monitoring strategy are depicted in Figure 3 below where the trends of wheel chamber expected and measured show an increasing difference the more the steam turbine is operating. After machine outage in September 2018 the two signals showed a similar behavior, this can be explained as a result of trapped steam condensation when machine stopped which resulted in soluble deposits dissolution in the steam path. The effect was the restoration of machine performances close to the expected and, consequently, the operation of the machine in a 'clean' condition for a limited amount of time.

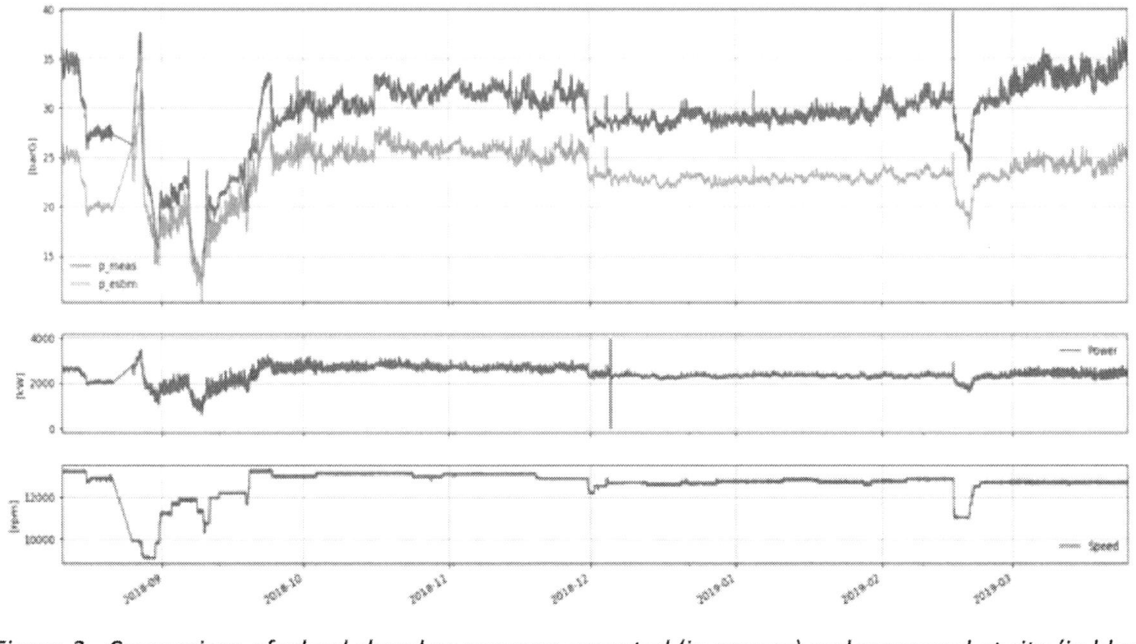

Figure 3 - Comparison of wheel chamber pressure expected (in orange) and measured at site (in blue)

A limit of 5% difference between the two signals was assumed to consider the machine as fouled, this is shown in Figure 4 where machine fouled data are the red dots. The graph of Figure 4 depicts the positioning of Steam Turbine (ST) clean data compared with ST fouled data. When the machine operates closer to rated conditions, namely at 2500 kW, in order to guarantee the rated power, the steam consumption is higher for the red dots showing ST fouled data. Thus, to guarantee a higher amount of steam flow, the control valve works with higher valve lifts, as presented in the lower part of the figure. The capability of combining these different sources of information gives a detailed overview on the steam turbine operation and represents also a great enabler for an efficient monitoring of the performances of the whole compression train.

The case-study described presents the trends of a backpressure steam turbine without extractions nor injections. Future developments will deal with extension of the analytic to other case-studies with different layouts and machine sizes.

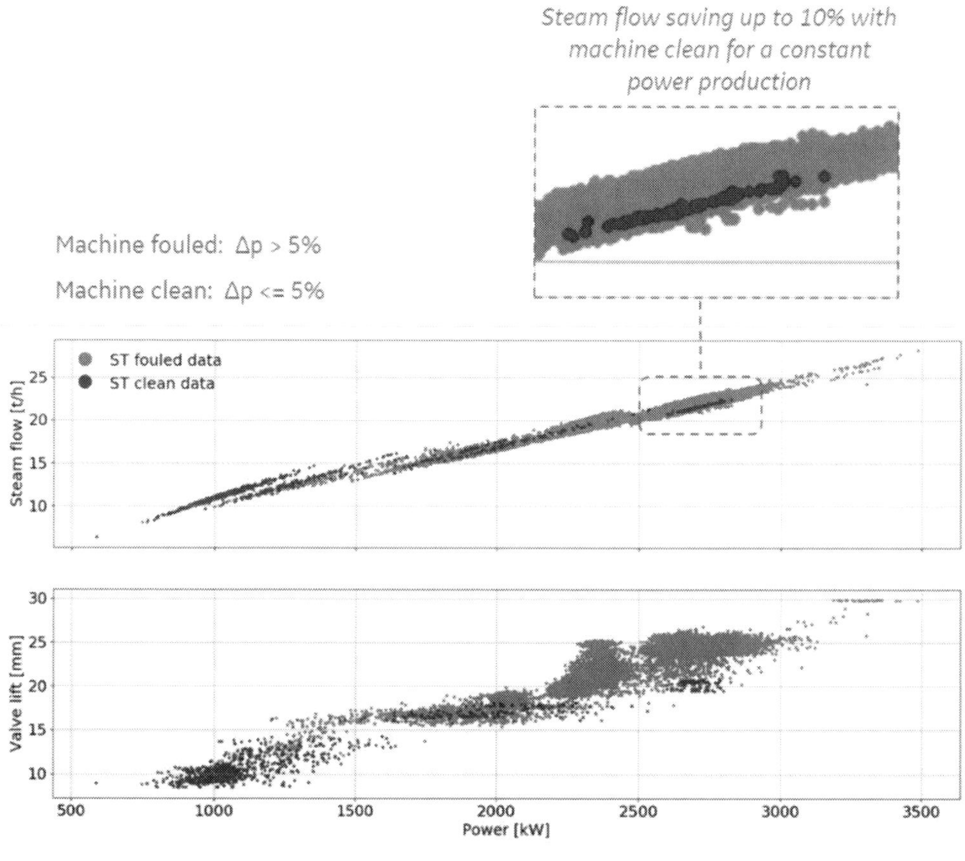

Figure 4 - Steam flow VS Power and Valve lift VS Power at machine fouled and clean conditions

5. Conclusions

The described case study shows the feasibility of modeling the fouling of a Steam Turbine as a function of changes introduced by fouling to the internals geometry.

While the case study execution has benefited of the knowledge of the internals geometry (i.e. fixed and rotating blades design and relevant clearances), the benefits of the approach can be generalised to any Steam Turbine design, leveraging the Stodola Law, in which the Stodola "constant" is modeled as a parameter and varies according to contaminants deposition along the steam path.

The long-term monitoring of this parameter can be applied to prevent severe machine damage and to suggest the proper timing for machine cleaning procedures.

Future developments will deal with the extension of the analytic to other case-studies with different layouts and machine sizes. In addition, based on the analytic, monitoring rules will be implemented to take into account also thrust bearing temperature, that is a further indication of increased pressures across the steam path.

In order to extend the benefits of the analytic, a dedicated study to continuously acquire and monitor the most common steam contaminants, and suggest proper water treatment actions will be completed.

References

1. World Economic Forum White paper (2017), "Digital Transformation Initiative - Chemistry and Advanced Materials Industry".

2. Gomez B., Snider B. (2013), "Successful Online and Offline Cleaning of Steam Turbines With and Without Disassembly", 42nd Turbomachinery and 29th Pump Symposia, Houston.

3. Zhou S. and Turnbull A. (2002), "Steam Turbine Operating Conditions, Chemistry of Condensates, and Environment Assisted Cracking – A Critical Review", NPL Report MATC (A) 95.

4. Wang L. and Zhang R. (2010), "A Detection Method for Steam Turbine Flow Path Fouling Based-on Support Vector Regression".

5. Tang G., Zhang G., Jianmingn G. (2010), "Fouling fault predict of steam turbine flow passage based on KPCA and LS-SVMR", Published in International Conference on Mechanic Automation.

551494 Impact of Time on Asphaltene Destabilization Detection in Unconventional Fuels

Much attention has been given to the time-dependent aspects of asphaltene destabilization and aggregation in crude oils in the recent years. It has been shown that many traditional asphaltene stability analyses, such as spot tests and titration techniques, fail to account for destabilization time and as a result they generate false negative results, i.e., conditions where asphaltene destabilization are not detected but will generate fouling in processing equipment and pipes. In this work, recent findings on asphaltene destabilization, aggregation, and deposition in crude oil will be presented. The impact of these findings on processing and use of unconventional fuels, bitumen, refined oils, and marine fuels will then be discussed.

HEAVY OIL REFORMING
138- Processing Unconventional Crudes/Alternative Fuels

Girish Srinivas*, Steve Schwab, Harsha Vempati, Brian Gebhard, Steven Gebhard

TDA Research, Inc.
4663 Table Mountain Drive
Golden, Colorado 80403, USA
(303) 940-2321
gsrinivas@tda.com

Abstract

TDA Research has developed a process to steam reform heavy hydrocarbon fractions such as atmospheric and vacuum residuum (also called atmospheric tower bottoms, ATB and vacuum tower bottoms, VTB) over nickel-based steam reforming catalysts without catalyst deactivation. The catalyst is not deactivated because it is continuously regenerated with air. In an industrial process with circulating fluidized beds (CFB), the catalyst would be continuously regenerated, but in our laboratory experiments, the catalyst was regenerated after about 90-120 minutes of steam reforming by first purging the reactor with nitrogen (to eliminate residual fuel) and then switching to an air feed. In this sense our process functions somewhat like a chemical looping steam reformer (CLR).

Introduction

Refineries that process heavy, sour crudes that contain metals, sulfur, and high molecular weight aromatic hydrocarbons require considerable hydrogen. Many sour crude oils originate in the Western Hemisphere, including heavy crudes from Venezuela, southern California, and oil sands in Canada. Processing and upgrading these heavy feedstocks requires considerable hydrogen, but unfortunately revamping or installing new hydrogen capacity with conventional technologies such as steam methane reforming (SMR) or petroleum coke gasification can be prohibitively expensive, especially in countries where natural gas prices are higher than they are in the U.S. (e.g. Japan, Germany and the UK which depend on liquefied natural gas imports, Figure 1).

Japan is the 4th largest refiner in the world and second largest in Asia with a capacity of about 4.25 Mbd and 2012 throughput of 3.4 Mbd. India is the 5th largest refiner in the world with a capacity of 4.1 Mbd and is the 4th largest oil consumer after the United States, China and Japan. India imports most of its crude from the Middle East. Given the high cost of liquefied natural gas (LNG) and the existing refining capacity in Japan, India, and the rest of Asia, TDA's technology could provide a lower cost way to make hydrogen than conventional steam methane reforming (SMR) where natural gas is reacted with steam at high temperatures in a fixed bed reactor. This is because TDA's process uses bottom of the barrel feedstocks that are available in all of these countries. TDA's technology converts heavy oil into syngas, which can be further shifted and purified to produce pure hydrogen. TDA's process uses steam reforming in a circulating fluidized bed reactor system that uses a catalytic heavy-hydrocarbon steam reformer and a catalyst regenerator.

TDA Research Inc. is developing a new, patent-pending (Allowed: Application No. 14/698,541, filed 04-28-2015, titled "Process for Generating Hydrogen from Heavy Oil or Hydrocarbons"), technology that will allow hydrogen to be produced in refineries at a cost that is considerably lower than hydrogen produced from conventional technologies (e.g. SMR) when natural gas prices are high or purchasing hydrogen from a third party. TDA's technology converts "bottom of the barrel" residuum and other heavy refinery feedstocks into hydrogen via a chemical looping steam reforming process (Figure 2).

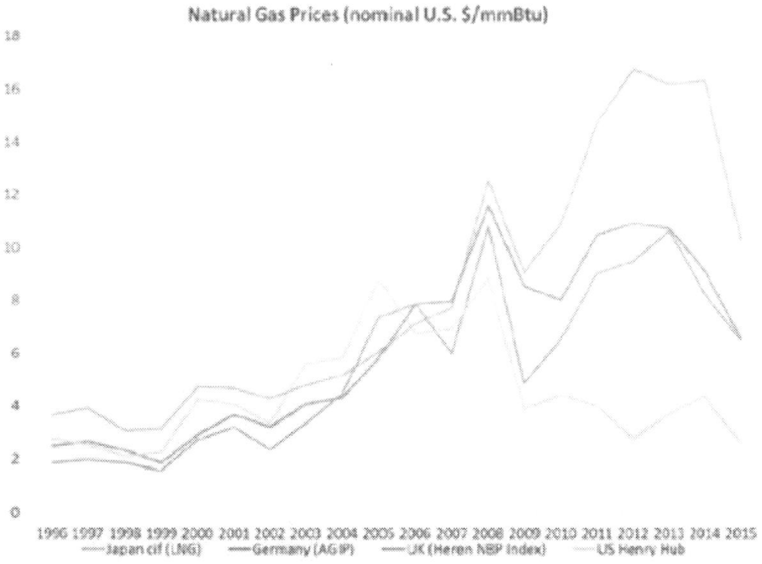

Figure 1. Historical international natural gas prices (State Utility Forecasting Group – March 2017).

In TDA's process, heavy oils are steam reformed over nickel-based catalysts to produce hydrogen without catalyst deactivation and without the need for an oxygen plant. This greatly expands the range of feedstocks that can be used to generate hydrogen. In the laboratory, we have steam reformed atmospheric and vacuum residuum [also known as atmospheric tower bottoms (ATB) and vacuum tower bottoms (VTB)] as well as lighter feedstocks such as vacuum gas oil (VGO) and naphtha over nickel based catalysts without catalyst deactivation. The catalyst does not deactivate because the process uses a fluidized bed

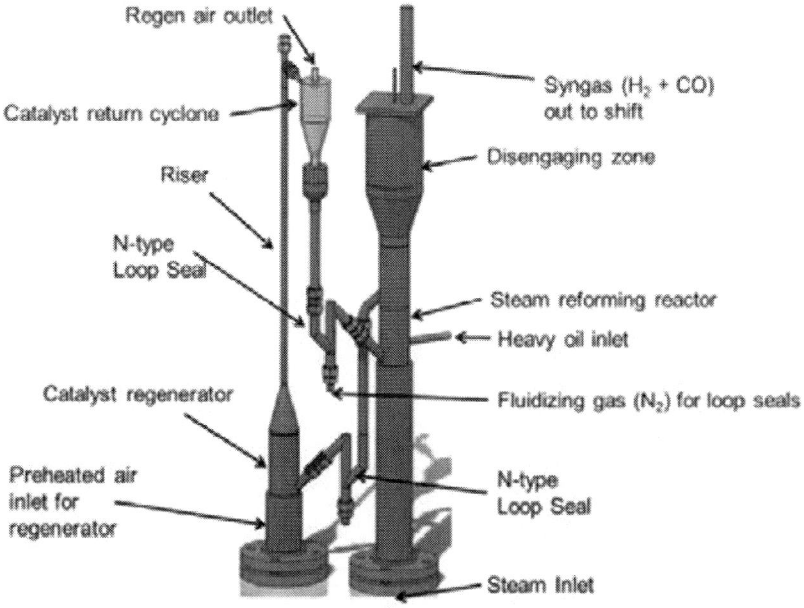

Figure 2. Preliminary design of a laboratory-scale, Incoloy 800H, CFB reactor system for heavy-oil steam reforming using TDA's process.

reformer along with continuous catalyst regeneration with air. The process is essentially chemical looping steam reforming (1,3). For example, residuum and steam are fed into a fluidized bed reactor containing Ni-based steam reforming catalyst at 850-900°C to generate syngas ($CO + H_2$). Because the process uses contact times on the order of minutes, not enough carbon builds up on the catalyst to cause irreversible deactivation. The catalyst is then regenerated by burning the coke in air at about 1000°C. In our laboratory scale reactor, a nitrogen purge was used between reforming and regeneration to prevent combustion; however, in an industrial setting, a circulating fluidized bed system would be used. Figure 2 is a SolidWorks drawing of a laboratory scale reactor that is being built at TDA using Incoloy 800HT.

Burning off the coke in the regenerator with air helps reheat the catalyst to about 1000°C for the next reforming step (a small amount of residuum, heavy oil or other fuel can be added to the regenerator to increase the temperature for additional heat). However, most of the heat is generated when Ni in the catalyst is oxidized to NiO, which is highly exothermic. The hot nickel catalyst returns to the reformer as NiO but is quickly reduced back to catalytically active nickel metal by the hydrocarbons in the feed and steam reforming resumes. The catalyst is continuously circulated between the two fluidized bed reactors.

TDA has conducted an extensive series of tests with atmospheric tower bottoms (ATB) and two types of vacuum tower bottoms (VTB) as well as VGO, naphtha, and diluted oil sand bitumen (dilbit) as feedstocks. Two VTB samples were tested, one heavy VTB and one lighter VTB (that we refer to as "medium" VTB). The medium VTB is liquid at room temperature and was simply fed to the reactor through a heated line using a high pressure ISCO syringe pump. The very heavy VTB had to be cut with 20% xylene to be pumpable with our laboratory equipment (the ISCO pump cannot operate at 150°C, which is required to liquefy the heavy VTB). With atmospheric tower bottoms (ATB), it was possible to operate with steam to carbon (S/C) ratios as low as 3 without catalyst deactivation. Both VTB feeds required operating with a steam/carbon ratio of 5. Significant amounts of coke deposited around the feed injector in the reactor at lower S/C ratios. No catalyst deactivation has been observed with ATB or either VTB sample. Figure 3 is a schematic of how the process

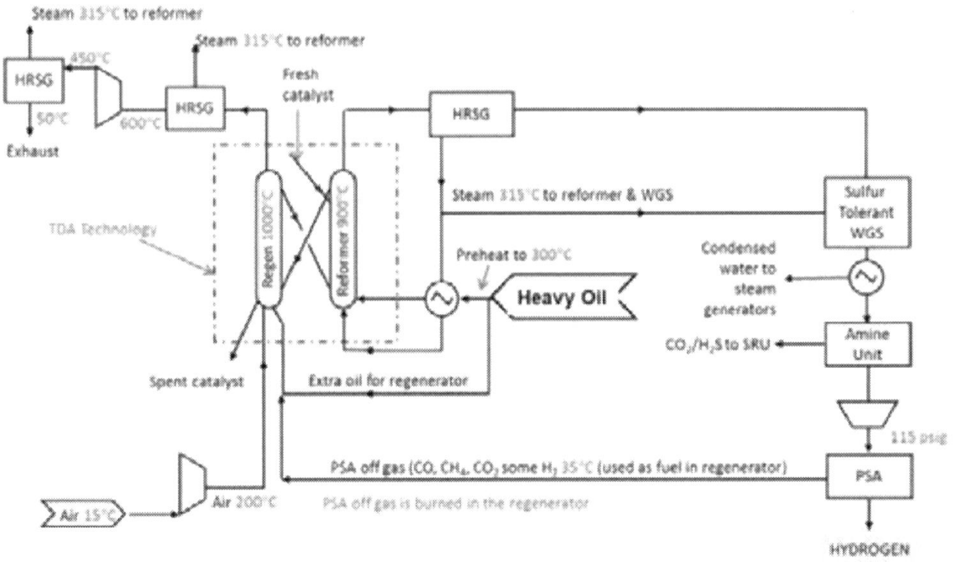

Figure 3. Schematic of TDA's heavy oil to hydrogen process.

would fit into a refinery, and Figure 4 is a conceptual drawing of the process that shows the chemical and catalytic reactions taking place in the reformer and regenerator.

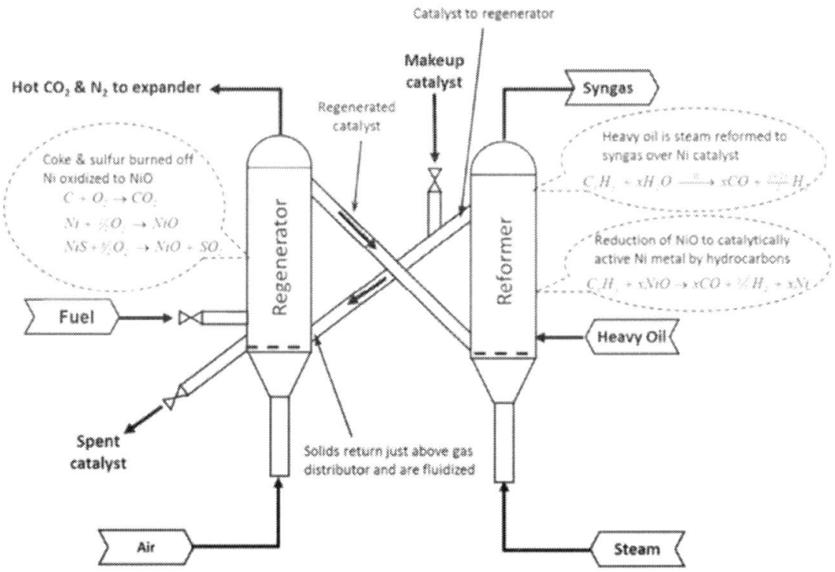

Figure 4. Conceptual schematic of TDA's process showing the reactions that occur during steam reforming and catalyst regeneration.

In the TDA process, heavy oil and steam are fed into a fluidized bed reactor containing a Ni-based steam reforming catalyst at 850-900°C to generate syngas (CO + H_2). Because reforming is carried out for only about 90-120 min before the catalyst is regenerated, the amount of carbon (coke) that builds up on the catalyst surface is not enough to cause irreversible deactivation. The catalyst is then sent to a regenerator where the coke (and any sulfur) is burned off with air. Coke is burned off in air, which partly reheats the catalyst for the next reforming cycle. Combustion of coke on the catalyst surface ($C + O_2 \rightarrow CO_2$, ΔH = -94 kcal/mole) and the oxidation of nickel on the catalyst to nickel oxide ($Ni + \frac{1}{2} O_2 \rightarrow NiO$, ΔH = -56 kcal/mole) are both exothermic with Ni oxidation suppling most of the energy required to increase the catalyst temperature to ~1000°C in the regenerator. Additional heavy hydrocarbon feed can be burned in the regenerator if needed to increase the catalyst temperature, but in general, Ni oxidation is sufficient. The hot catalyst that returns to the reforming reactor is now in the form of nickel oxide, which is catalytically inactive for steam reforming; however, in the our process, the hydrocarbons in the feed quickly reduce the NiO back to catalytically active Ni for the next reforming cycle ($xNiO + C_xH_y \rightarrow xCO + y/2\ H_2 + xNi$). In an industrial sized unit, the sensible heat in the catalyst provides most of the energy required by the endothermic steam-hydrocarbon reforming reactions. The catalyst circulation rate, fuel feed, steam, and regenerator flow rates are adjusted to maintain the proper heat balance in the process.

The economics of hydrogen generation in refineries strongly depends on feedstock cost. The price of natural gas for industrial consumers in the U.S. is currently low (Figure 1) making steam methane reforming (SMR) economically attractive for generating hydrogen. The selling price of "bottom of the barrel" products depends on the price of petroleum coke and high sulfur, heavy fuel oil. Nevertheless, increasing gas prices, in countries where natural gas prices are high or in countries that lack of SMR infrastructure (e.g. smaller refineries), TDA's hydrogen from residuum process becomes economically attractive.

Another area where our residuum steam reforming process may be useful is for generating hydrogen for upgrading oil sand bitumen during syncrude production. Currently, syncrude is made by coking raw oil sand bitumen, which converts about 40-50 wt% of the carbon in the bitumen into coke. The liquid syncrude is then piped to refineries. Because coking the bitumen removes what would otherwise be the bottoms fraction of the crude, the syncrude is usually blended with conventional petroleum so that it can be processed in existing refineries. TDA has demonstrated that our process can use oil sand bitumen to produce hydrogen. In the laboratory, we tested bitumen diluted with condensate (dilbit) since this is a liquid at room temperature; in industrial settings raw bitumen could be heated to become fluid. By using a fraction of the mined bitumen to produce hydrogen, it may be possible to either reduce the amount of coking needed to produce syncrude, hydrotreat the bitumen directly, or upgrade syncrude produced by conventional methods before it enters the pipeline, which would increase the yield of high quality syncrude.

TDA's heavy oil steam reforming process's main competitors are partial oxidation (POX) and heavy oil gasification. In contrast to these technologies, our process benefits from the advantages of oxygen gasification *without the need for an oxygen separation plant*, and we operate at much lower temperature than gasification. Because we use a separate vessel for catalyst regeneration, little if any nitrogen is introduced into the syngas, which improves the efficiency of downstream water gas shift (WGS) and hydrogen separation operations such as pressure swing adsorption (PSA). All of the downstream processes shown in Figure 3 (WGS, PSA, etc.) are proven technologies. A preliminary economic analysis suggests that we can produce hydrogen for approximately \$6/1000 SCF using our process compared with ~\$9/1000 SCF from a small (0.5-10 MMSCFD) steam methane reforming (SMR) plant, or purchasing delivered liquid hydrogen for ~\$9-28/1000 SCF.

Experimental Methods

The laboratory experiments were done using the apparatus shown in Figure 5. The apparatus uses a single fluidized bed, which is cycled between heavy oil steam reforming and catalyst regeneration. The reactor is briefly purged with nitrogen between steps to prevent mixing air with heavy oil/syngas. Electronic mass flow controllers are used to meter five gas streams into the reactor via ¼"- 316 stainless steel tubing. Each gas stream

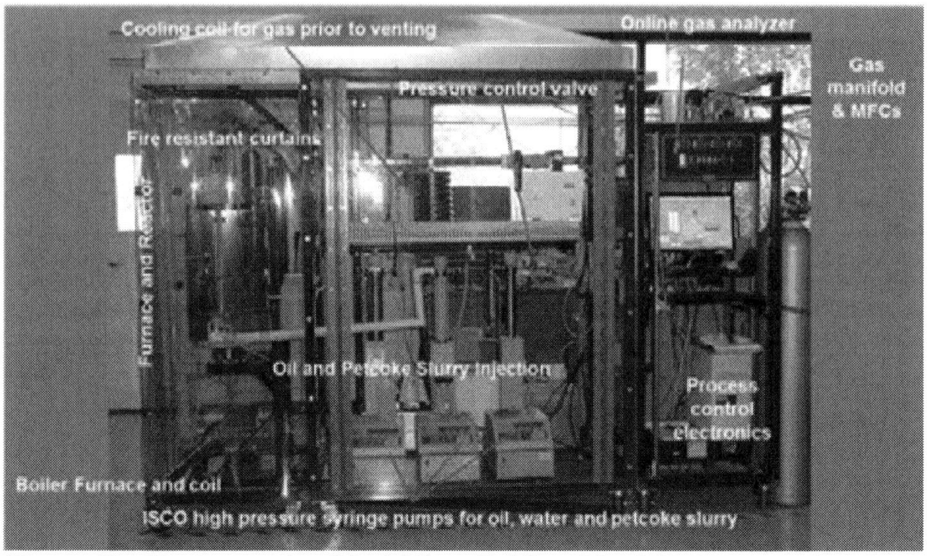

Figure 5. Photograph of bubbling fluidized bed test apparatus.

can be isolated by means of pneumatically actuated valves. There is also a high pressure liquid pump (ISCO stainless steel "syringe pump") that is used to feed water to a boiler (a tubing coil in a temperature-controlled furnace) where it is vaporized into steam. A second ISCO pump is used to feed heavy oil to a water-cooled injector where a nitrogen sweep aids in oil delivery. The residuum feedstocks tested were atmospheric tower bottoms (ATB) and two samples of vacuum tower bottoms (VTB) obtained from different refineries; we have also demonstrated the process using vacuum gas oil (VGO - both heavy and light) and diluted oil sand bitumen. The oil pump has a heated jacket filled with silicone oil at 70-90°C that is circulated by a heated oil bath to keep the viscosity of the oil low enough that it can be pumped.

Because of the high temperatures (850-900°C), the fluidized bed reactor is made from SCH40 Incoloy 800H pipe. A section of 4" pipe is used as a disengaging zone, which is welded to a 4"x 2" bell reducer that is welded to a section of 2" pipe that contains the catalyst bed. The catalyst bed contains about 500g of a Ni-based steam reforming catalyst. Downstream from the reactor, a 10 μm filter collects any catalyst dust that is entrained in the gas flow, a heat exchanger/condensing coil to remove water, and a sodium citrate bubbler/scrubber to prevent H_2S (during reforming) and SO_2 (during regeneration) from entering the online H_2, CO, CO_2, CH_4 gas analyzer (Nova Analytical Systems). There is also a separate online paramagnetic O_2 analyzer to measure breakthrough during catalyst regeneration. The system pressure is controlled using a computer-controlled, pressure control valve (PCV) made by Badger Meter Co. A computer (PC) running OPTO22 software controls the PCV as well as the heating tapes, tube furnace, mass flow controllers, and other components of the system.

Results & Discussion

Tests with Atmospheric Tower Bottoms

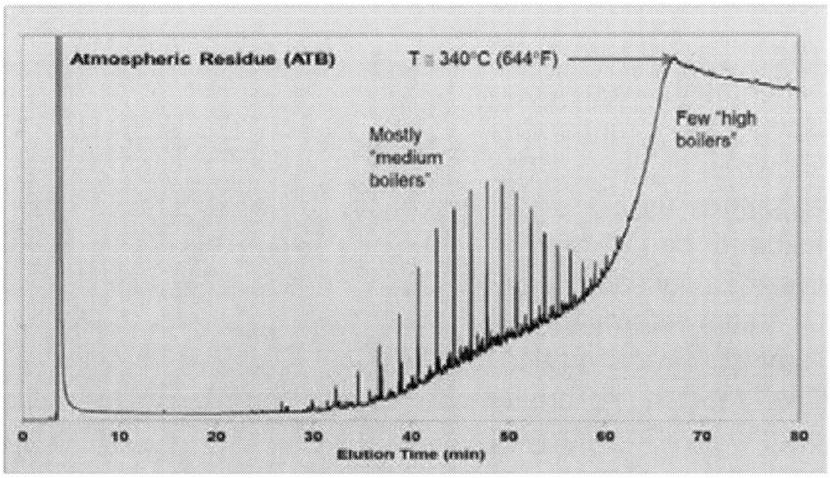

Figure 6. Gas chromatogram of ATB.

The first heavy oil feedstock we tested was atmospheric tower bottoms (ATB, also called long residuum) from a working refinery. A capillary column gas chromatogram of the oil dissolved at a concentration of 10 mg/mL in dichloromethane is shown in Figure 6. The baseline increases due to GC column bleed at high temperatures (the temperature at the highest point in Figure 6 occurs at T = 340°C (644°F). Even though the ATB stream is typically 650+°F material, small amounts of volatile hydrocarbons are still present that elute from the GC column between 30 and 60 min, which corresponds to about 250-600°F [roughly kerosene through gas oil, (2)].

The steam reforming experiments were run with a Ni-based catalyst and the steam to carbon ratios were varied between 1.5 and 13.6. Steam reforming and catalyst regeneration were done at 865°C, and 900°C, respectively, at a total system pressure of 50 psig. Depending on the test, steam reforming lasted 90-120 minutes, and catalyst regeneration lasted 40-60 min, so one complete cycle was about 3 hours long (e.g. ~25 cycles in ~80 hours).

Figure 7 shows the product gas composition measured using the online gas analyzers for a single steam reforming and catalyst regeneration cycle using ATB. In all cases the carbon balances were 85-95% (within experimental error). The most important result is that steam reforming ATB does not require excessively high S/C ratios. In fact, after extensive testing, the optimum steam/carbon ratio for ATB was found to be about 3 and for VTB was about 5, which is encouraging because this is in the range of S/C ratios used industrially for steam naphtha reforming (5).

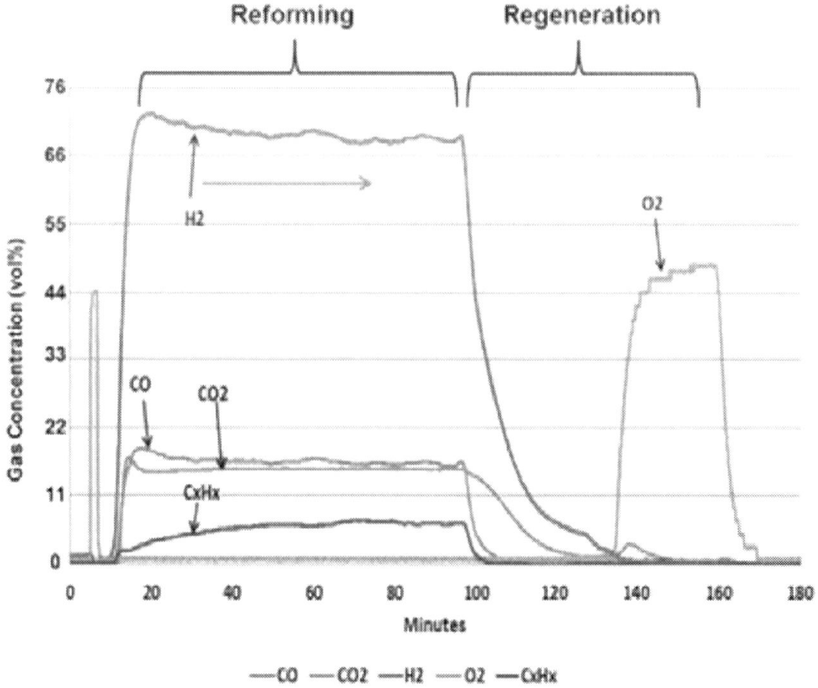

Figure 7. Single cycle of ATB reforming with S/C = 3 using Ni-based catalyst at 50 psig and 865°C.

Figure 8 shows the results for approximately 80 hours of cycling between ATB steam reforming and catalyst regeneration. Because the maximum concentration of the red line (H_2) returns to the same level with each cycle, there was no catalyst deactivtion over the course of the test. If permanent catalyst deactivation were

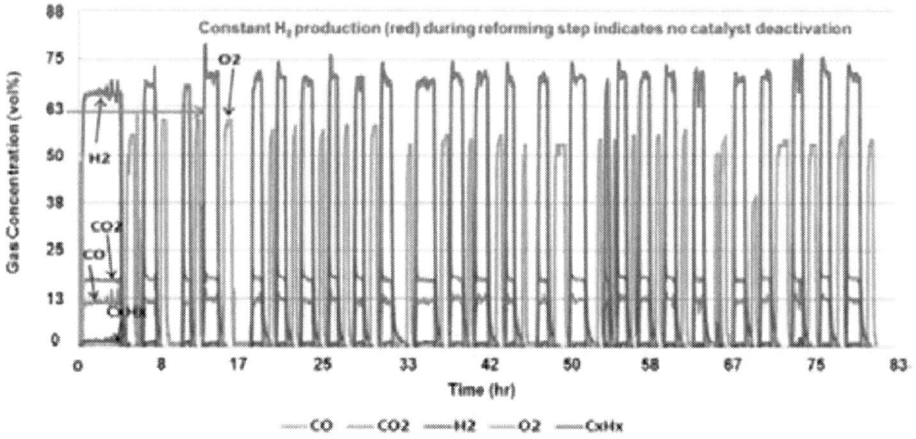

Figure 8. Gas composition for steam reforming ATB at 865°C, 50 psig, S/C = 5, Ni-based catalyst.

occuring, there would be downward trend with time on stream as less hydrogen would be produced during each cycle. The concentration of hydrogen in the gas remained relatively constant at 70-75 vol% (dry basis).

Tests with "Medium" Vacuum Residuum

Our success with ATB encouraged us to try a heavier but more economically attractive feed – vacuum residuum (a.k.a. vacuum tower bottoms – VTB). The VTB samples came from a working refinery. Figure 9 is a gas chromatogram for the "medium" VTB. Its' GC analysis was surprisingly similar to the ATB in that kerosine through gas oil hydrocarbons were present (to about the same extent). Some heavier components appear as well. The main reason that the heavier component peaks are not larger (as might be expected) is that the "medium" VTB is less soluble in dichloromethane (CH_2Cl_2, the solvent used for GC analysis) than ATB. Nevertheless, comparing the gas chromatograms of the ATB and two VTB samples is qualitatively useful for characterizing their compositional differences.

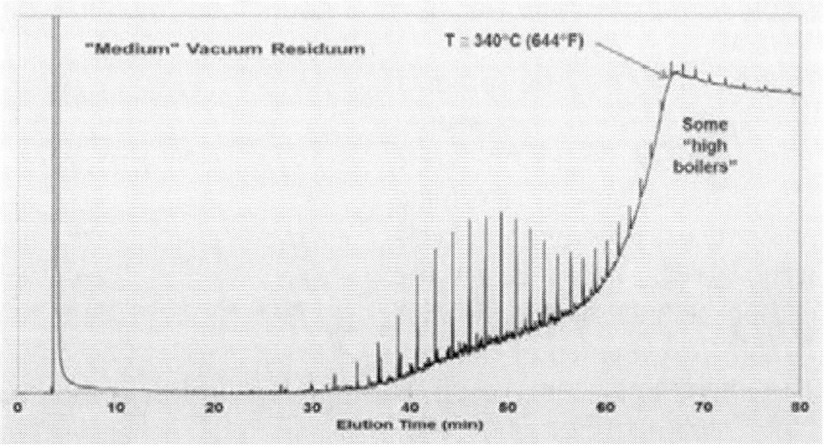

Figure 9. Gas chromatogram of "medium" vacuum residuum.

Figure 10 shows the results of about 275 hours of steam reforming-catalyst regeneration cycles with the "medium" VTB. As was the case with the ATB, no deactivation of the Ni catalyst was observed over the course of the 275 hours, indicating that coke and sulfur deposited on the catalyst were burned off in the regenerator and that than any metals (e.g. Ni and V present in residuum) had no measurable effect. As before, the hydrogen concentration was about 70 vol% (dry basis).

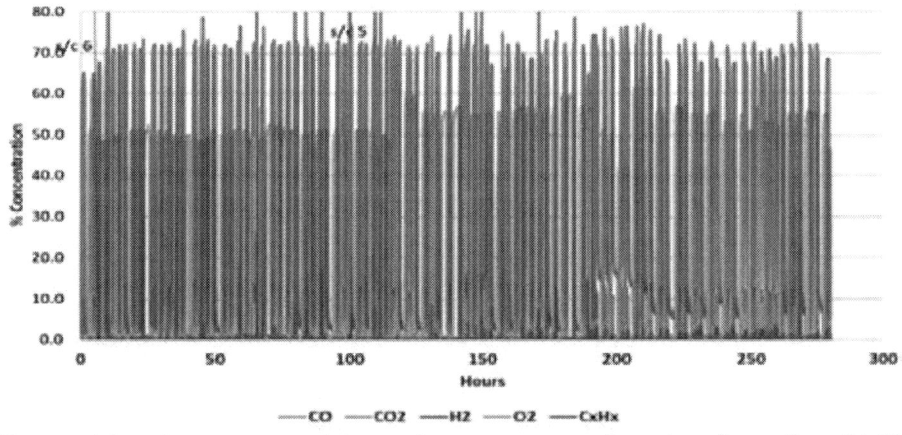

Figure 10. Gas composition when steam reforming "medium" VTB at 865°C, 50 psig, S/C =5, Ni-based catalyst.

Tests with "Heavy" Vacuum Residuum

Figure 11 is a gas chromatogram for the "heavy" VTB. This material was a solid at room temperature and must be heated to almost 150°C to melt. Unfortunately, this temperature is too high to be used in our liquid feed system (that uses an ISCO syringe pump) so we were forced to dilute it with 20 wt% xylene (mixed isomers). This mixture still required some heating to be pumpable.

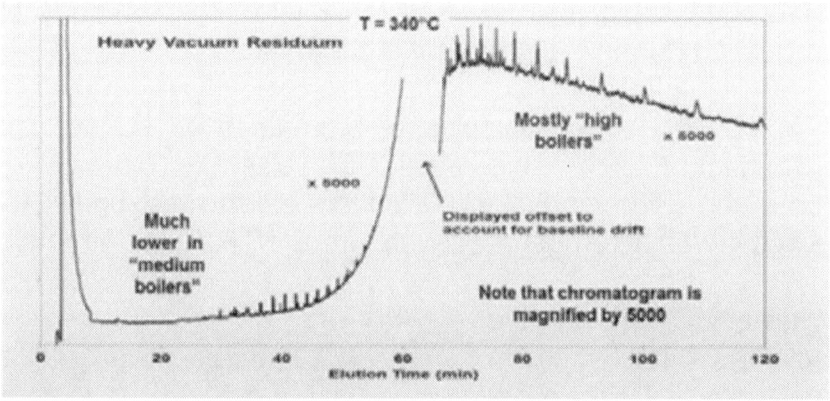

Figure 11. Gas chromatogram of "heavy" vacuum residuum.

The "heavy" VTB was much more "tar-like" than the "medium" VTB and had very little in the way of kerosine through gas oil hydrocarbons. Heavier components are present to a larger extent than in the "medium" VTB in spite of the fact that this material has even lower solubility in dichloromethane. The GC peaks have been enlarged by a factor of 5000 to make them more visible. The offset is where the background (column bleed) rose to very high levels. Cutting out the time between 60 and 70 min (where there were no peaks anyway) allows the high temperature data to be shown on the same graph. This "heavy" VTB is only slightly soluble in dichloromethane (CH_2Cl_2) so the peaks in the raw GC data are quite small. Nevertheless, the qualitative comparison between Figure 11 and Figure 9 shows that this VTB is considerably heavier than the "medium" VTB and therefore contains greater amounts of high molecular weight hydrocarbons.

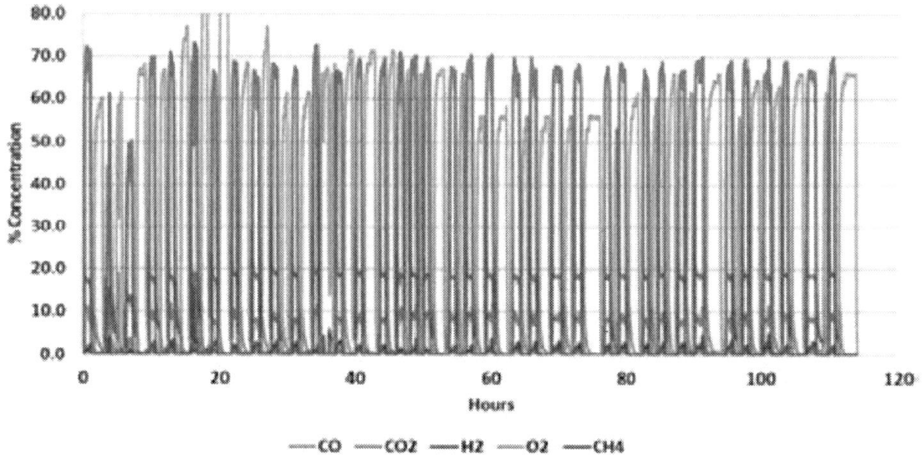

Figure 12. Gas composition when steam reforming "heavy" VTB at 865°C, 50 psig, S/C =5, Ni-based catalyst.

Figure 12 shows the results of about 115 hours of steam reforming-regeneration cycles with the "heavy" VTB. As before, no deactivation of the Ni catalyst was observed over the course of the run, and the hydrogen concentration was about 70 vol% (dry basis). We have also done tests with vacuum gas oil (VGO: both light and heavy) and obtained similar results.

Tests with Diluted Oil Sand Bitumen

Canadian Oil Sands (bitumen) are an important unconventional petroleum resource. Normally, bitumen is coked to produce a synthetic crude oil (missing the 1000+°F fraction). The syncrude is then pipelined to refineries (4). Syncrude (which has no bottoms due to the coking step) must be blended with conventional oil so that the atmospheric and vacuum distillation equipment can function properly. One alternative or adjunct to coking to produce syncrude from bitumen would be to use some of the raw bitumen (or DILBIT) to make hydrogen for upgrading bitumen to syncrude to reduce the amount of coking required. Hydrogen could also be used to stabilize the syncrude by saturating olefins and aromatics produced during the coking operation.

We obtained three samples of bitumen from Sunoco that were classified as: 1) DILBIT (Bitumen + 30% condensate), 2) Sales Oil (Bitumen + 15% condensate) and finally 3) Emulsion (Bitumen/water). We were able to test both DILBIT and Sales Oil but not Emulsion since it was heterogeneous with lumps of insoluble bitumen in water.

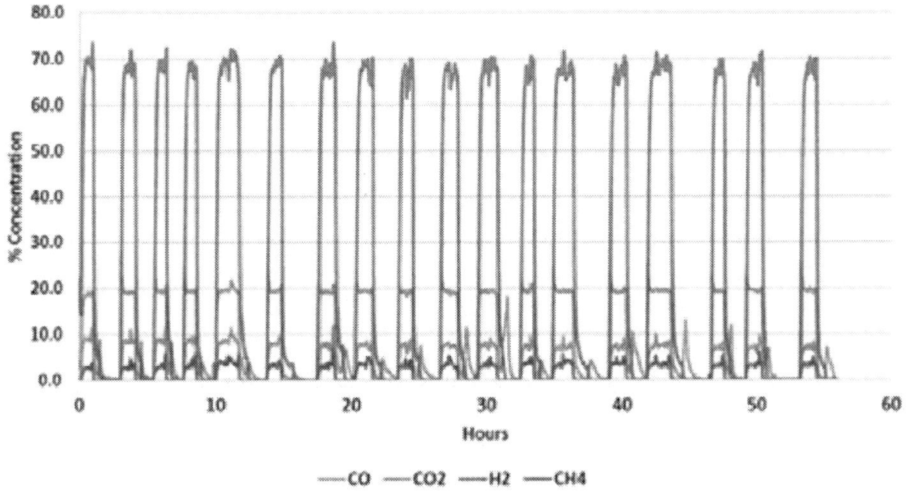

Figure 13. Gas composition when steam reforming diluted oil sand bitumen 865°C, 50 psig, S/C =5, Ni-based catalyst.

Figure 13 shows the results for steam reforming DILBIT in our laboratory reactor. The condensate used to dilute the bitumen is lighter liquid hydrocarbons that are produced along with bitumen during Steam Assisted Gravity Drainage (SAGD) since some in-situ steam distillation of the material naturally occurs underground. DILBIT is liquid at room temperature and was easy to feed to our steam reforming reactor. The performance of TDA's process with DILBIT was essentially identical to that obtained with refinery residuum ATB and VTB (i.e. 70 vol% H_2 in raw syngas and no catalyst deactivation).

Discussion and Conclusions

The reason that the Ni-based, steam reforming catalyst does not deactivate in TDA's process is that cycling between steam reforming and catalyst regeneration removes coke and sulfur from the catalyst before it can build up to damaging levels. Hydrocarbons are steam reformed to syngas in the reformer reactor, and coke and sulfur that deposit on the catalyst are burned off in the regenerator. Nickel is also oxidized to NiO in the regenerator, which heats the catalyst to approximately 1000°C. Hot catalyst from the regenerator (now NiO) is reduced back to catalytically active Ni metal by the hydrocarbons in the feed when it returns to the steam reformer. The sensible heat of the regenerated catalyst supplies the energy required by the endothermic steam reforming reactions. In this sense, not only is the Ni a steam reforming catalyst, but as importantly a heat transfer material. In a continuous system, balancing the circulation rate between the reformer and regenerator along with appropriate flow rates of heavy oil and steam, allows for the continuous production of syngas. The syngas is then water-gas-shifted (WGS) downstream to increase the H_2 concentration, followed by H_2 purification using pressure swing adsorption (PSA) or a membrane system.

In conclusion, TDA has successfully demonstrated that we can steam reform a wide variety of heavy oil feedstocks to generate hydrogen for refineries. Our economic calculations indicate that the process becomes competitive when the natural gas price approaches $6/MM Btu ($6/1000 SCF). These results show that unlike fixed bed steam reforming, TDA's process to produce hydrogen from heavy oil can be used with feedstocks as heavy as vacuum residuum because the process uses continuous catalyst regeneration. This makes TDA's process very feedstock flexible because virtually any liquid hydrocarbon stream can be used.

References

1. Adanez, J.; Abad, A.; Garcia-Labiano, F.; Gayan, P. and de Diego, L.F. (2012) Review: Progress in Chemical-Looping Combustion and Reforming Technologies," *Progress in Energy and Combustion Science* **38** (2012) 215-282
2. Gary, J.H. and Handwerk, G.E. (2001) Petroleum Refining, Technology and Economics, 4th ed., Marcel Dekker, Inc.
3. Hoteit, A.; Forret, A.: Pelletant, W.: Roesler, J. and Gauthier, T. (2011) "Chemical Looping Combustion with Different Types of Liquid Fuels," *Oil & Gas Science and Technology* - Rev. IPP Energies Nouvelles, **66** (2) pp. 193-199
4. Probstein, R.F. and Hicks, R.E. (1982) Synthetic Fuels, McGraw Hill
5. Twigg, M.V. (1996) Catalyst Handbook, 2nd ed., Ch. 5, "Steam Reforming," Manson Publishing

550389 Producing Renewable Distillate Products in Refineries

Refiners face both opportunities and challenges when processing alternative feedstocks. By leveraging existing equipment and infrastructure, capital costs can be minimized and existing assets can be fully utilized, but only if unique processing challenges are addressed. Both stand-alone and co-processing flow sheets will be reviewed along with options for retrofit of existing hydroprocessing units.

550590 Development of Thermodynamic Parameters for Amine-Containing H2S Scavengers

Removal of H_2S from oil and gas streams has been a priority, oftentimes a necessity, for upstream crude oil/gas facilities for decades. However, the advances in production, namely horizontal drilling and hydraulic fracking, have produced oils and natural gases that contain higher concentrations of H_2S than previously seen. These sour crudes must undergo separation and/or conversion processes to lower the H_2S content prior to entering the midstream pipelines and railcars to prevent equipment corrosion/fouling and to reduce human exposure. H_2S scavengers can be categorized into regenerative and non-regenerative types. Regenerative, including alkanolamines, are used in typical absorption and stripping recovery processes and are betters suited for larger wellhead facilities. Small well head facilities, on the other hand, benefit from non-regenerative scavengers, including acrolein, formaldehyde, glyoxal, and triazines, that react with H2S and incorporate the sulfur into a hydrocarbon molecule. Triazine-based scavengers have become the most widely used of the reactive-type scavengers. 1,3,5-hexahydrotriazine is the most frequently used non-regenerative scavenger; however, its byproducts are deleterious to downstream equipment (i.e. overhead condensers) due to basic conditions (pH 8).

These harmful effects have led to molecular adaptations to the basic triazine molecule whose byproducts would be benign to equipment fouling and corrosion. Varying side chains have the impact of making the triazine scavenger more (or less) soluble in aqueous (or non-polar crudes). In addition, these side chains could be tailored to be more active for sour gas, rather than sour liquid crude, streams. Side chain modifications lead to uncertainty of solubility, liquid/gas interactions, and possible byproduct complications. Since scavengers such as melamine and cyanuric acid have been studied extensively, this study focuses on developing thermodynamic parameters (i.e. solubility, scavenger/ion interations, etc.) to for implementation into activity coefficient models for triazines containing ethanolic, amino, bromine, and acrolein substituents, including tris (2,3-dibromopropyl) isocyanurate, 1,3,5-Triacryloylhexahydro-1,3,5-triazine, and 1,3,5-Tris (2-hydroxyethyl) cyanuric acid . Using potentiometric titration and GC-MS analyzes, the solubility of these compounds in water and in saltwater solutions (using different NaCl concentrations) has been evaluated. The solubility results have yielded insightful information towards the ionic mediums in which these materials could be used. Moreover, the octanol/water partition coefficients have been determined to examine the potential fate these scavengers considering environmental factors. The experimental results and model development will be presented.

550503 Estimation of Fouling Threshold Parameters for Crude Oil Heat Exchanger Networks Using Data Reconciliation and Gross Error Detection

In crude oil refineries, fouling in heat exchangers and heat exchanger networks (HENs) represents a challenging and complicated issue. Its occurrence produces major impacts in thermal, hydraulic and economic performance. Increase in fuel consumption, pressure drop across heat exchangers and CO_2 emissions are examples of a wide series of consequences that are attributed to fouling in almost every crude oil refinery. In order to mitigate fouling, different approaches have been developed. Previous studies have discovered correlations among fouling deposition and different operational variables such as wall temperature and flow velocity. This dependency has been further exploited for defining several techniques for either design or retrofit of heat exchangers to mitigate the impact of fouling. Current approaches aim to obtain an accurate simulation model for heat exchanger networks, as well as methodologies for including fouling deposition. If succeeded, these simulation models can potentially improve fouling predictions and decision-making processes related to operational optimisation and cleaning schedules. The most common fouling models available are threshold models. These models are able to define a set of operating conditions in which fouling can be avoided. However, threshold models present a series of parameters that need to be adjusted for each type of crude oil, affecting the design of generalised strategies for fouling modelling in crude oil heat exchanger networks. This set of parameters can be estimated via process measurements as these realistically represent the interactions within a HEN. The main challenges are the presence of measurement error and the amount of available data. Both of these challenges can be accounted for using data reconciliation and gross error detection. The adjustment of measurements in order to satisfy process constraints and the identification of faulty measurements are complementary methods that allow for obtaining reliable data and for a deep analysis of the existing plant-instrumentation. This work proposes a flexible-structured simulation scheme for heat exchanger networks, along with the minimisation of the measurement error, and a parameter estimation based on stochastic optimisation for calculating fouling model parameters. The use of this methodology allows for simultaneous calculation of fouling model parameters for each side of heat exchangers within a HEN. Moreover, the data reconciliation and gross error detection algorithms assess the quality of the measured data, and at the same time analyses the estimation-potential of unmeasured variables, whenever possible. All these insights are shown in a case study, where a crude oil HEN subject to fouling is used to test the proposed methodology. Results show that the use of data reconciliation improves the reliability of measurements and faulty instruments are correctly identified, allowing for an accurate estimation of fouling model parameters. The results from the parameter estimation can be used in assessing the thermal performance of a HEN, as well as in the prediction of fouling behaviour, which is paramount for the design and optimisation of fouling mitigation strategies.

550540 Application of Plate Heat Exchangers into Crude Oil Preheat Trains

Heat exchanger enhancement technologies have been widely used into heat exchanger networks (HENs) retrofit of crude oil preheat train. However, the heat recovery is limited through the enhancement, which is restricted by the geometry of existing heat exchanger. Plate heat exchangers are widely used in energy-intensive process industries to increase heat recovery and reduce greenhouse gas emissions with the features of small minimum temperature approach and high effectiveness. Thus, plate heat exchangers are taken into consideration to replace conventional heat exchanger in crude oil preheat trains to enhance heat transfer coefficient and reduce energy consumption. A major limitation of applying plate heat exchanger is lack of reliable optimization method to exploit benefit effectively. Due to the high installation cost of plate heat exchangers, the optimization is based on the trade-off between energy reduction and capital cost. This work proposes a new methodology of application of plate heat exchanger in HENs retrofit with a fixed network structure. However, energy saving is limited without structure modification. Therefore, this work also presents a novel pinch method with the application of plate heat exchangers to provide cost effective retrofit with structure modification. Case studies highlight the benefits of the new approaches.

550111 Utilizing Smart Meter Verification for Greenhouse Gas Emissions Reporting

Government greenhouse gas (GHG) requirements state that utility providers and industrial users producing GHG emissions from burning fossil fuels or from chemical reactions, must measure and report fuel oil and natural gas burned or the GHG emissions from combustion and report other sources of GHG emissions. One key regulation requires sources emitting greater than 25,000 metric tons of CO_2 or the equivalent (this threshold varies by location) to measure either the emissions or fuel burned and report quarterly. There are a few different methods refiners and petrochemical facilities can utilize to report on emissions. In this session we will compare and evaluate the different methods and how diagnostic information from instrumentation can be used to ensure accuracy of reporting and ensure compliance. We will also explore additional benefits associated with different fuel measurement and control options for combustion applications.

551928 Soft Skills for Young Professionals

Going from being a student to working is a big transition for most of us. It requires us to learn how to navigate the real world, and we can use all the help we can get to do it well. Soft skills are the most important secondary skill (the primary one being the vocation that you are being employed for) that can help us here, and not only in our careers but also in personal lives. This session will focus on some of the necessary soft skills regardless of the degree we have, reflect on which soft skills are transferrable and which need to be developed in order to have a rewarding time at our organizations.

551786 The *Trusted Advisor*: A Model for Career Success

The classic book *Trusted Advisor* by David H. Maister presents highly effective tactics for building and maintaining long lasting relationships as part of a successful career strategy. A Trusted Advisor is a person who has purposefully elevated their personal and professional relationships to a level where solving problems is more valuable than technical mastery. Trusted Advisors have self-confidence and driven curiosity. They believe that there is distinction between business life and personal life, but both are seamlessly interconnected.

This presentation will outline the three core skills needed to become a Trusted Advisor: earning trust, giving advice and building relationships. Effective ways to develop each of these three skills will be presented in an easy-to-understand format. Pitfalls for destroying trust will be discussed along with tactics to avoid doing so. The goal of the presentation is to outline these tactics in the form of constructive actions that you can perform daily for the rest of your career. Trusted Advisors are easy to spot: they seem confident, selfless and laser focused on helping themselves and others achieve their goals.

550124 Networking Strategies to Propel Your Career

This presentation discusses various early career strategies to build business relationships with coworkers, clients, and vendors. Develop a strong network now for any future downturn or advancing within your company.

550521 Engineering Creative Thinking

What do you think of when you hear the word Creativity? If you're like me, you often think of something artsy that you would see on Instagram or Pinterest. You also probably think creativity is a trait people are born with or without. You probably also think creativity is something you don't have to utilize in your technical jobs. However, creativity and creative thinking is a skill that can set you apart in any job. Creativity isn't about originality or thinking outside of the box as much as it is about learning how to connect ideas, growing a bigger box, and building your mental inventory of knowledge. In this session we will go over practices, tools, and tricks for how us engineers can expand our creative thinking to make us more innovative and strategic in how we approach our problem solving and our jobs.

AUTHOR INDEX

AIChE
120 Wall Street, FL 23
New York, NY 10005-4020

ISBN 978-1-5108-8677-3

CURRAN ASSOCIATES INC.
proceedings.com